# Smithells Light Metals Handbook

# Smithells Light Metals Handbook

Edited by
**E. A. Brandes** *CEng, BSc(Lond), ARCS, FIM*
and
**G. B. Brook** *DMet(Sheff), FEng, FIM*

Butterworth-Heinemann
Linacre House, Jordan Hill, Oxford OX2 8DP
225 Wildwood Avenue, Woburn, MA 01801-2041
A division of Reed Educational and Professional Publishing Ltd

℞ A member of the Reed Elsevier plc group

OXFORD   BOSTON   JOHANNESBURG
MELBOURNE   NEW DELHI   SINGAPORE

First published 1998

© Reed Educational and Professional Publishing Ltd 1998

**British Library Cataloguing in Publication Data**
A catalogue record for this book is available from the British Library

ISBN 0 7506 3625 4

**Library of Congress Cataloguing in Publication Data**
A catalogue record for this book is available from the Library of Congress

Typeset by Laser Words, Madras, India
Printed and bound in Great Britain

# Contents

# Preface

The light metals covered by this handbook are only those of industrial importance – aluminium, magnesium and titanium. The values given have been updated to the time of publication. They are intended for all those working with light metals; for research or design purposes Reference to source material may be found in *Smithells Metals Reference Book* (revised 7th edition). For design purpose values of mechanical properties must be obtained from the relevant specifications. Equilibrium diagrams are taken to be the most suitable for general work. For specialist work on any system, original sources should be consulted.

*E.A.B.*
*Chalfont St Peter, Bucks*

# 1 Related specifications

**Table 1.1** RELATED SPECIFICATIONS FOR WROUGHT ALUMINIUM ALLOYS

| BS international | Nominal composition old ISO No. Al– | UK former BS designation | France former NF | W. Germany Wk. No. | Canada | Sweden | USSR | Italy — Old UNI | Italy — New UNI | Japan |
|---|---|---|---|---|---|---|---|---|---|---|
| 1050A | 99.5 | 1B | A5 | 3.0255 | – | 4007 | – | 4507 | 9001/2 | A1050 |
| 1080A | 99.8 | 1A | A8 | 3.0285 | – | 4004 | – | 4509 | 9001/4 | A1080 |
| 1200 | 99 | 1C | A4 | 3.0205 | 990 | 4010 | – | 3567 | 9001/1 | A1200 |
| 1350 | 99.5 | 1E | A5/6 | 3.0257 | – | – | – | – | 9001/5 | – |
| 2011 | Cu6BiPb | FC1 | A–U5 PbBi | 3.1655 | CB60 | 4355 | – | 6362 | 9002/5 | A2011 |
| 2014A | Cu4SiMg | H15 | A–U4SG | 3.1255 | CS41N | 4338 | AK 8 | 3581 | 9002/3 | A2014 |
| 2017A | Cu4MgSi | – | A–U4G | 3.1325 | – | – | – | – | – | – |
| 2024 | Cu4Mg1 | 2L97, 2L98, L109, L110, DTD5100A | A–U4G1 | 3.1355 | CG42 | – | D 16 | 3583 | 9002/4 | A2024 |
| 2031 | Cu2NiMgFeSi | H12 | A–U2N | – | CG30 | – | D 18 | 3577 | 9002/1 | A2217 |
| 2117 | Cu2Mg | 3L86 | A–U2G | 3.1305 | – | – | – | 3578 | 9002/6 | – |
| 2618A | Cu2Mg1.5Fe1Ni1 | H16 | A–U2GN | – | – | – | AK 4-1 | – | – | – |
| 3103 | Mn1 | N3 | – | 3.0515 | – | 4054 | – | 3568 | 9003/3 | – |
| 3105 | MnMg | N31 | – | 3.0505 | – | – | – | – | 9003/5 | A3105 |
| 4043 | Si5 | N21 | A–S5 | – | – | – | – | – | – | A4043 |
| 4047 | Si12 | N2 | A–S12 | – | – | – | – | – | – | – |
| 5005 | Mg1 | N41 | A–G0.6 | – | – | 4106 | – | 5764 | 9005/1 | A5005 |
| 5056A | Mg5 | N6 | A–G5M | 3.3555 | – | – | – | – | – | – |
| 5083 | Mg4.5Mn | N8 | A–G4.5MC | 3.3547 | GM41 | 4140 | – | 7790 | 9005/5 | A5083 |
| 5154A | Mg3.5 | N5 | – | – | GR40 | – | AMG3 | – | – | – |
| 5251 | Mg2 | N4 | A–G2M | 3.3525 | GM31N | – | – | 3574 | 9005/3 | – |
| 5454 | Mg3.6 | N51 | A–G2.5MC | 3.3537 | GM31P | – | – | 7789 | – | A5454 |
| 5554 | Mg3Mn | N52 | – | – | GS11N | – | – | – | 9006/2 | A5554 |
| 6061 | MgSiCu | H20 | A–GSUC | 3.3211 | GS10 | 4104 | AD3 | 6170 | – | A6061 |
| 6063 | Mg0.5Si | H9 | – | – | – | – | AD31 | – | – | A6063 |
| 6082 | Si1MgMn | H30 | A–SGM0.7 | 3.2315 | – | 4212 | – | 3571 | 9006/4 | – |
| 7020 | Zn4.5Mg | H17 | A–Z5G | 3.4335 | – | 4425 | – | 7791 | 9007/1 | – |
| 7075 | Zn6MgCu | 2L95, L160, L161, L162 | A–Z5GU | 3.4365 | ZG62 | SM6958 | V95 | 3735 | 9007/2 | A7075 |

**Table 1.2** RELATED SPECIFICATIONS FOR MAGNESIUM ALLOYS

Cast alloys

| Nominal composition | UK designation | UK BS2970 MAG | USA ASTM | USA AMS | France AFNOR | Standard AECMA | W. Germany aircraft | W. Germany DIN 1729 |
|---|---|---|---|---|---|---|---|---|
| RE3Zn2.5 Zr0.6 | ZRE1 | 6-TE | EZ33A-T5 | 4442B | G-TR3Z2 | MG-C-91 | 3.6204 | 3.5103 |
| Zn4.2RE1.3 Zr0.7 | RZ5 | 5-TE | ZE41A-T5 | 4439A | G-Z4TR | MG-C-43 | 3.6104 | 3.5101 |
| Th3Zn2.2 Zr0.7 | ZT1 | 8-TE | HZ32A-T5 | 4447B | G-Th3Z2 | MG-C-81 | 3.6254 | 3.5105 |
| Zn5.5Th1.8 Zr0.7 | TZ6 | 9-TE | ZH62A-T5 | 4438B | – | MG-C-41 | 3.5114 | 3.5102 |
| Al8Zn0.5 Mn0.3 | A8 | 1-M | AZ81A-F | – | G-A9 | MG-C-61 | – | 3.5812 |
| Al9.5Zn0.5 Mn0.3 | AZ91 | 3-7B | AZ91C-T4 | – | G-A9Z1 | – | 3.5194 | – |
| Al7.5/9.5 Zn0.3/1.5 Mn0.15min | C | 7-M | – | – | – | – | – | 3.5912 |

1.4.2 Wrought alloys

| | | | | | | | | |
|---|---|---|---|---|---|---|---|---|
| Zn3Zr0.6 | ZW3 | E-151M | – | – | – | MG-P-43 | – | – |
| Al6Zn1Mn0.3 | AZM | E-121M | AZ61A-F | 4350H | G-A6Z1 | MG-P-63 | W.3510 | 3.5612 |
| Al8.5Zn0.5 Mn0.12min | AZ80 | – | AZ80A | 4360D | – | MG-P-61 | W.3515 | 3.5812 |
| Al3Zn1Mn0.3 | AZ31 | S-1110 | AZ31B-0 | 4375F | G-A2Z1 | MG-P-62 | W.3504 | 3.5312 |

**Table 1.3** TITANIUM AND TITANIUM ALLOYS. CORRESPONDING GRADES OR SPECIFICATIONS

| IMI designation | UK British Standards (Aerospace series) and Min. of Def. DTD series* | France AIR-9182, 9183, 9184 | Germany BWB series† | AECMA recommendations | USA AMS series‡ | USA ASTM series |
|---|---|---|---|---|---|---|
| IMI 115 | BS TA 1, DTD 5013 | T-35 | 3.7024 | Ti-POI | | ASTM grade 1 |
| IMI 125 | BS TA 2,3,4,5 | T-40 | 3.7034 | Ti-PO2 | AMS 4902, 4941, 4942, 4951 | ASTM grade 2 |
| IMI 130 | DTD 5023, 5273, 5283, 5293 | T-50 | | | AMS 4900 | ASTM grade 3 |
| IMI 155 | BS TA 6 | T-60 | 3.7064 | Ti-PO4 | { AMS 4901 / AMS 4921 | ASTM grade 4 |
| IMI 160 | BS TA 7,8,9 | | | | | |
| IMI 230 | BS TA 21, 22, 23, 24, BS TA 52–55, 58 | T-U2. | 3.7124 | Ti-P11 | | |
| IMI 315 | DTD 5043 | | | | | |
| IMI 317 | BS TA 14, 15, 16, 17 | T-A5E | | Ti-P65 | AMS 4909, 4910, 4926, 4924, 4953, 4966 | |
| IMI 318 | BS TA 10, 11, 12, 13, 28, 56 | T-A6V | 3.7164 | Ti-P63 | AMS 4911, 4928, 4934, 4935, 4954, 4965, 4967 | ASTM grade 5 |
| IMI 318 ELI (extra low interstitial) | – | T-A6VELI | | | AMS 4907, 4930, 4931 | ASTM grade 3, F. 136. |

*continued overleaf*

**Table 1.3**    (*continued*)

| IMI designation | UK British Standards (Aerospace series) and Min. of Def. DTD series* | France AIR-9182, 9183, 9184 | Germany BWB series† | AECMA recom- mendations | USA AMS series‡ | USA ASTM series |
|---|---|---|---|---|---|---|
| IMI 325 | – | T-A3V2.5 | 3.7194 | | AMS 4943 4944 | ASTM grade 9 |
| IMI 550 | BS TA 45–51, 57 | T-A4DE | 3.7184 | Ti-P68 | | |
| IMI 551 | BS TA 39–42 | | | | | |
| IMI 624 | – | T-A6Zr4 DE | 3.7144 | | AMS 4919, 4975, 4976 | |
| IMI 646 | – | – | – | – | AMS 4981 | |
| IMI 662 | – | – | 3.7174 | – | AMS 4918, 4971, 4978, 4979 | |
| IMI 679 | BS TA 18–20, 25-27 | | | | AMS 4974 | |
| IMI 680 | DTD 5213 | T-E11DA | | | | |
| IMI 685 | BS TA 43, 44 | T-A6ZD | 3.7154 | Ti-P67 | | |
| IMI 811 | – | T-A8DV | | | AMS 4915, 4916 | |
| IMI 834 | – | T-A6E Zr4Nb | | | | |

*UK BS 3531 Part 1 (Metal Implants in Bone Surgery), and Draft British Standard for Lining of Vessels and Equipment for Chemical Processes, Part 9, also refer.

†Germany DIN 17850, 17860, 17862, 17863, 17864 (3.7025/35/55/65), and TUV 230-1-68 Group I, II, III and IV also refer.

‡USA MIL-T-9011, 9046, 9047, 14577, 46038, 46077, 05-10737 and ASTM B265-69, B338-65, B348-59T, B367-61T. B381-61T, B382-61T also refer.

# 2 General physical properties of light metal alloys and pure light metals

## 2.1 General physical properties of pure light metals and their alloys

**Table 2.1** PHYSICAL PROPERTIES OF ALUMINIUM, MAGNESIUM AND TITANIUM

| Property | Aluminium | Magnesium | Titanium |
|---|---|---|---|
| Atomic weight C = 12 | 26.98154 | 24.305 | 47.88 |
| Atomic number | 13 | 12 | 22 |
| Density (g cm$^{-1}$) | 2.70 | 1.74 | 4.5 |
| liquid at 660 °C Al, 651 °C Mg, 1685 °C Ti | 2.385 | 1.590 | 4.11 |
| Melting point °C | 660.323 (fixed pt ITS-90) | 649 | 1667 |
| Boiling point °C | 2520 | 1090 | 3285 |
| Thermal conductivity Wm$^{-1}$K$^{-1}$ at °C | | | |
| 0–100 | 238 | 155.5 | 16 |
| 200 | 238 | 167 | 15 |
| 400 | 238 | 130 | 14 |
| 600 | – | – | 13 |
| 800 | – | – | (13) |
| Specific heat Jkg$^{-1}$K$^{-1}$ at °C | | | |
| 20 | 900 | 1022 | 519 |
| 0–100 | 917 | 1038 | 528 |
| 200 | 984 | 1110 | 569 |
| 300 | 1030 | – | – |
| 400 | 1076 | 1197 | 619 |
| 600 | – | – | 636 |
| 800 | – | – | 682 |
| Coefficient of expansion 10$^{-6}$K$^{-1}$ at °C | | | |
| 0–100 | 23.5 | 26.0 | 8.9 |
| 100 | 23.9 | 26.1 | 8.8 |
| 200 | 24.3 | 27.0 | 9.1 |
| 300 | 25.3 | – | – |
| 400 | 26.49 | 28.9 | 9.4 |
| 600 | – | – | 9.7 |
| 800 | – | – | 9.9 |
| Electrical resistivity $\mu\Omega$ at °C | | | |
| 20 | 2.67 | 4.2 | 54 |
| 100 | 3.55 | 5.6 | 70 |
| 200 | 4.78 | 7.2 | 88 |
| 300 | 5.99 | – | – |
| 400 | 7.30 | 12.1 | 119 |
| 600 | – | – | 152 |
| 800 | – | – | 165 |
| Temp. coefficient of resistivity 0–100 °C 10$^{-3}$K$^{-1}$ | 4.5 | 4.25 | 3.8 |

Note: For surface tension and viscosity of liquid metals see *Metal Reference Book* 7th ed. pp. 14–7 to 14–8

## 2.2   The physical properties of aluminium and aluminium alloys

**Table 2.2**   THE PHYSICAL PROPERTIES OF ALUMINIUM AND ALUMINIUM ALLOYS AT NORMAL TEMPERATURES
*sand cast*

| Material | Nominal composition % | | Density g cm$^{-3}$ | Coefficient of expansion 20–100°C 10$^{-6}$ K$^{-1}$ | Thermal conductivity 100°C W m$^{-1}$ K$^{-1}$ | Resistivity μΩ m | Modulus of elasticity MPa ×10$^3$ |
|---|---|---|---|---|---|---|---|
| Al | Al | 99.5 | 2.70 | 24.0 | 218 | 3.0 | 69 |
| | Al | 99.0 | 2.70 | 24.0 | 209 | 3.1 | – |
| Al–Cu | Cu | 4.5 | 2.75 | 22.5 | 180 | 3.6 | 71 |
| | Cu | 8 | 2.83 | 22.5 | 138 | 4.7 | – |
| | Cu | 12 | 2.93 | 22.5 | 130 | 4.9 | – |
| Al–Mg | Mg | 3.75 | 2.66 | 22.0 | 134 | 5.1 | – |
| | Mg | 5 | 2.65 | 23.0 | 130 | 5.6 | – |
| | Mg | 10 | 2.57 | 25.0 | 88 | 8.6 | 71 |
| Al–Si | Si | 5 | 2.67 | 21.0 | 159 | 4.1 | 71 |
| | Si | 11.5 | 2.65 | 20.0 | 142 | 4.6 | – |
| Al–Si–Cu | Si | 10 | 2.74 | 20.0 | 100 | 6.6 | 71 |
| | Cu | 1.5 | | | | | |
| | Si | 4.5 | 2.76 | 21.0 | 134 | 4.9 | 71 |
| | Cu | 3 | | | | | |
| Al–Si–Cu–Mg* | Si | 17 | 2.73 | 18.0 | 134 | 8.6 | 88 |
| | Cu | 4.5 | | | | | |
| | Mg | 0.5 | | | | | |
| Al–Cu–Mg–Ni | Cu | 4 | 2.78 | 22.5 | 126 | 5.2 | 71 |
| (Yalloy) | Mg | 1.5 | | | | | |
| | Ni | 2 | | | | | |
| Al–Cu–Fe–Mg | Cu | 10 | 2.88 | 22.0 | 138 | 4.7 | 71 |
| | Fe | 1.25 | | | | | |
| | Mg | 0.25 | | | | | |
| Al–Si–Cu–Mg–Ni | Si | 12 | 2.71 | 19.0 | 121 | 5.3 | 71 |
| (Lo–Ex) | Cu | 1 | | | | | |
| | Mg | 1 | | | | | |
| | Ni | 2 | | | | | |
| | Si | 23 | 2.65 | 16.5 | 107 | – | 88 |
| | Cu | 1 | | | | | |
| | Mg | 1 | | | | | |
| | Ni | 1 | | | | | |

*Die cast.

**Table 2.3**   THE PHYSICAL PROPERTIES OF ALUMINIUM AND ALUMINIUM ALLOYS AT NORMAL TEMPERATURES

*wrought*

| Specification | Nominal composition % | | Condition* | | Density g cm$^{-3}$ | Coefficient of expansion 20–100°C 10$^{-6}$ K$^{-1}$ | Thermal conductivity 100°C W m$^{-1}$ K$^{-1}$ | Resistivity μΩ cm | Temp. coeff. of resistance 20–100°C | Modulus of elasticity MPa ×10$^3$ |
|---|---|---|---|---|---|---|---|---|---|---|
| 1199 | Al | 99.992 | Sheet | H111 | 2.70 | 23.5 | 239 | 2.68 | 0.0042 | 69 |
|  |  |  |  | H18 |  |  | 234 | 2.70 | 0.0042 | 69 |
| 1080A | Al | 99.8 | Extruded |  | 2.70 | 23.5 | 239 | 2.68 | 0.0042 | 69 |
|  |  |  | Sheet | H111 |  |  | 234 | 2.74 | 0.0042 | 69 |
|  |  |  |  | H18 |  |  | 230 | 2.76 | 0.0042 | 69 |
| 1050A | Al | 99.5 | Extruded |  | 2.71 | 23.5 | 230 | 2.79 | 0.0041 | 69 |
|  |  |  | Sheet | H111 |  |  | 230 | 2.80 | 0.0041 | 69 |
|  |  |  |  | H18 |  |  | 230 | 2.80 | 0.0041 | 69 |
| 1200 | Al | 99 | Extruded |  | 2.71 | 23.5 | 226 | 2.82 | 0.0041 | 69 |
|  |  |  | Sheet | H111 |  |  | 226 | 2.85 | 0.0041 | 69 |
|  |  |  |  | H18 |  |  | 226 | 2.87 | 0.0040 | 69 |
|  |  |  | Extruded |  |  |  | 226 | 2.89 | 0.0040 | 69 |
|  |  |  |  |  |  |  | 226 | 2.86 | 0.0040 | 69 |
| 2014A | Cu | 4.4 | Extruded | T4 | 2.8 | 22 | 142 | 5.3 |  |  |
|  | Mg | 0.7 |  | T6 | 2.8 | 22 | 159 | 4.5 |  | 74 |
|  | Si | 0.8 |  |  |  |  |  |  |  |  |
|  | Mn | 0.75 |  |  |  |  |  |  |  |  |
| 2024 | Cu | 4.5 |  | T3 | 2.77 | 23 |  | 5.7 |  | 73 |
|  | Mg | 1.5 |  | T6 | 2.77 | 23 | 151 | 5.7 |  | 73 |
|  | Mn | 0.6 |  |  |  |  |  |  |  |  |
| 2090 | Cu | 2.7 |  | T8 | 2.59 | 23.6 | 88.2 | 9.59 |  | 76 |
|  | Li | 2.3 |  |  |  |  |  |  |  |  |
|  | Zr | 0.12 |  |  |  |  |  |  |  |  |
| 2091 | Cu | 2.1 |  | T8 | 2.58 | 23.9 | 84 | 9.59 |  | 75 |
|  | Li | 2.0 |  |  |  |  |  |  |  |  |
|  | Mg | 1.50 |  |  |  |  |  |  |  |  |
|  | Zr | 0.1 |  |  |  |  |  |  |  |  |
| 3103 | Mn | 1.25 | Sheet | H111 | 2.74 | 23.0 | 180 | 3.9 | 0.0030 | 69 |
|  |  |  |  | H12 |  |  |  |  |  |  |
|  |  |  |  | H14 |  |  |  |  |  |  |
|  |  |  |  | H16 |  |  |  |  |  |  |
|  |  |  |  | H18 |  |  |  |  |  |  |
|  |  |  | Extruded |  |  |  | 151 | 4.8 | 0.0024 | – |

*continued overleaf*

**Table 2.3**  *(continued)*

*wrought*

| Specification | Nominal composition % | | Condition* | Density $g\,cm^{-3}$ | Coefficient of expansion 20–100°C $10^{-6}\,K^{-1}$ | Thermal conductivity 100°C $W\,m^{-1}\,K^{-1}$ | Resistivity $\mu\Omega\,cm$ | Temp. coeff. of resistance 20–100°C | Modulus of elasticity $MPa \times 10^3$ |
|---|---|---|---|---|---|---|---|---|---|
| 5083 | Mg | 4.5 | Sheet H111 | 2.67 | 24.5 | 109 | 6.1 | 0.0019 | 71 |
| | Mn | 0.7 | H12 | | | | | | |
| | Cr | 0.15 | H14 | | | | | | |
| 5251 | Mg | 2.0 | Sheet H111 | 2.69 | 24 | 155 | 4.7 | 0.0025 | 70 |
| | Mn | 0.3 | H13 | | | | | | |
| | | | H16 | | | | | | |
| 5154A | Mg | 3.5 | Extruded H111 | 2.67 | 23.5 | 147 | 4.9 | 0.0023 | – |
| | | | Sheet H14 | | | 142 | 5.3 | 0.0021 | 70 |
| 5454 | Mg | 2.7 | Extruded H111 | 2.68 | 24 | 138 | 5.4 | 0.0021 | – |
| | Mn | 0.75 | Sheet H22 | | | 134 | 5.7 | – | – |
| | Cr | 0.12 | H24 | | | 147 | 5.1 | 0.0019 | 70 |
| Al–Li | Li | 2.0 | Sheet T6 | 2.56 | – | – | – | – | 77 |
| Al–Mg–Li | Mg | 3.0 | Sheet T6 | 2.52 | – | – | – | – | 79 |
| Al–Li–Mg | Li | 2.0 | Sheet T6 | 2.46 | – | – | – | – | 84 |
| | Li | 3.0 | | | | | | | |
| | Mg | 2.0 | | | | | | | |
| 6061 | Mg | 1.0 | Bar H111 | 2.7 | 23.6 | 180 | | | 68.9 |
| | Si | 0.6 | T4 | 2.7 | 23.6 | 154 | | | 68.9 |
| | Cu | 0.2 | T6 | 2.7 | 23.6 | 167 | | | 68.9 |
| | Cr | 0.25 | | | | | | | |
| 6063 | Mg | 0.5 | Extruded T4 | 2.70 | 23.0 | 193 | 3.5 | 0.0033 | 71 |
| | Si | 0.5 | T6 | | | 201 | 3.3 | 0.0035 | – |
| 6063A | Mg | 0.5 | Bar T4 | 2.7 | 24 | 197 | 3.5 | | 69 |
| | Si | 0.5 | T5 | | 24 | 209 | 3.2 | | 69 |
| | | | T6 | | 24 | 201 | 3.3 | | 69 |
| 6082 | Mg | 1.0 | Bar/Extruded T4 | 2.7 | 23 | 172 | 4.1 | 0.0031 | 69 |
| | Si | 1.0 | T6 | 2.7 | 23 | 184 | 3.7 | 0.0031 | 69 |
| | Mn | 0.7 | | | | | | | |
| 6082 | Mg | 1.0 | Sheet T4 | 2.69 | 23.0 | 188 | 3.6 | 0.0033 | 69 |
| | Si | 1.0 | T6 | | | 193 | 3.4 | 0.0035 | – |

| Alloy | Composition | | Form | Temper | | | | | | |
|---|---|---|---|---|---|---|---|---|---|---|
| 6463 | Mg | 0.65 | Bar | T5 | 2.71 | 23.4 | 209 | 3.1 | | 69 |
| | Si | 0.4 | | T6 | 2.71 | 23.4 | 201 | 3.3 | | 69 |
| Al–Cu–Mg–Si (Duralumin) | Cu | 4.0 | Sheet | T6 | 2.80 | 22.5 | 147 | 5.0 | 0.0023 | 73 |
| | Mg | 0.6 | | | | | | | | |
| | Si | 0.4 | | | | | | | | |
| | Mn | 0.6 | | | | | | | | |
| | Cu | 4.5 | Sheet | T4 | 2.81 | 22.5 | 147 | 5.2 | 0.0022 | 73 |
| | Mg | 0.5 | | T6 | | | 159 | 4.5 | 0.0026 | – |
| | Si | 0.75 | | | | | | | | |
| | Mn | 0.75 | | | | | | | | |
| Al–Cu–Mg–Ni (Yalloy) | Cu | 4.0 | Forgings | T6 | 2.78 | 22.5 | 151 | 4.9 | 0.0023 | 72 |
| | Mg | 1.5 | | | | | | | | |
| | Ni | 2.0 | | | | | | | | |
| Al–Si–Cu–Mg (Lo–Ex) | Si | 12.0 | Forgings | T6 | 2.66 | 19.5 | 151 | 4.9 | 0.0023 | 79 |
| | Cu | 1.0 | | | | | | | | |
| | Mg | 1.0 | | | | | | | | |
| | Ni | 1.0 | | | | | | | | |
| Al–Zn–Mg | Zn | 10.0 | Forgings | | 2.91 | 23.5 | 151 | 4.9 | 0.0023 | – |
| | Cu | 1.0 | | | | | | | | |
| | Mn | 0.7 | | | | | | | | |
| | Mg | 0.4 | | | | | | | | |
| 7075 | Zn | 5.7 | Extrusion | T6 | 2.80 | 23.5 | 130 | 5.7 | 0.0020 | 72 |
| | Mg | 2.6 | | | | | | | | |
| | Cu | 1.6 | | | | | | | | |
| | Cr | 0.25 | | | | | | | | |
| 8090 | Li | 2.5 | Plate | | 2.55 | 21.4 | 93.5 | 9.59 | | 77 |
| | Cu | 1.3 | | | | | | | | |
| | Mg | 0.95 | | | | | | | | |
| | Zr | 0.1 | | | | | | | | |

H111 = Annealed.
H12,22 = Quarter hard.
H14,24 = Half hard.
H16,26 = Three-quarters hard.
H18,28 = Hard.

T4 = Solution treated and naturally aged.
T6 = Solution treated and artificially aged.

## 2.3  The physical properties of magnesium and magnesium alloys

**Table 2.4**  THE PHYSICAL PROPERTIES OF SOME MAGNESIUM AND MAGNESIUM ALLOYS AT NORMAL TEMPERATURE

| Material | Nominal composition [†] % | Condition | Density at 20°C g cm⁻³ | Melting point °C Sol. | Melting point °C Liq. | Coeff. of thermal expansion 20–200°C $10^{-6}\,K^{-1}$ | Thermal conductivity $W\,m^{-1}\,K^{-1}$ | Electrical resistivity μΩ cm | Specific heat 20–200°C $J\,kg^{-1}\,K^{-1}$ | Weldability by argon arc process [‡] | Relative damping capacity [§] |
|---|---|---|---|---|---|---|---|---|---|---|---|
| Pure Mag | Mg 99.97 | T1 | 1.74 | | 650 | 27.0 | 167 | 3.9 | 1050 | A | |
| Mg–Mn | (MN70)Mn 0.75 approx. | T1 | 1.75 | 650 | 651 | 26.9 | 146 | 5 | 1050 | A | |
| | (AM503)Mn 1.5 | T1 | 1.76 | 650 | 651 | 26.9 | 142 | 5.0 | 1050 | A | C |
| Mg–Al | AL80Al 0.75 approx. Be 0.005 | T1 | 1.75 | 630 | 640 | 26.5 | 117 | 6 | 1050 | A | |
| Mg–Al–Zn | (AZ31)Al 3 Zn 1 | T1 | 1.78 | 575 | 630 | 26.0 | (84) | 10.0 | 1050 | A | |
| | (A8)Al 8 Zn 0.5 | AC | 1.81 | 475* | 600 | 27.2 | 84 | 13.4 | 1000 | A | C |
| | | AC T4 | 1.81 | | | 27.2 | 84 | – | 1000 | | |
| | (AZ91)Al 9.5 Zn 0.5 | AC | 1.83 | 470* | 595 | 27.0 | 84 | 14.1 | 1000 | A | C |
| | | AC T4 | 1.83 | | | 27.0 | 84 | – | 1000 | | |
| | | AC T6 | 1.83 | | | 27.0 | 84 | – | 1000 | | |
| | (AZM)Al 6 Zn 1 | T1 | 1.80 | 510 | 610 | 27.3 | 79 | 14.3 | 14 000 | A | |
| | (AZ855)Al 8 Zn 0.5 | T1 | 1.80 | 475* | 600 | 27.2 | 79 | 14.3 | 1000 | A | |
| Mg–Zn–Mn | (ZM21)Zn 2 Mn 1 | T1 | 1.78 | | | 27.0 | | | – | A | |
| Mg–Zn–Zr | (ZW1)Zn 1.3 Zr 0.6 | T1 | 1.80 | 625 | 645 | 27.0 | 134 | 5.3 | 1000 | A | A |
| | (ZW3)Zn 3 Zr 0.6 | T1 | 1.80 | 600 | 635 | 27.0 | 125 | 5.5 | 960 | C | |
| | (25Z)Zn 4.5 Zr 0.7 | AC T6 | 1.81 | 560 | 640 | 27.3 | 113 | 6.6 | 960 | C | |
| | (ZW6)Zn 5.5 Zr 0.6 | T5 | 1.83 | 530 | 630 | 26.0 | 117 | 6.0 | 1050 | C | |

| Alloy system | Designation / composition | Condition | | | | | | | | | |
|---|---|---|---|---|---|---|---|---|---|---|---|
| Mg–Y–RE–Zr | (WE43) Y 4.0 / RE(Δ) 3.4 / Zr 0.6 | AC T6 | 1.84 | 550 | 640 | 26.7 | 51 | 14.8 | 966 | A | |
| | (WE54) Y 5.1 / RE(Δ) 3.0 / Zr 0.6 | AC T6 | 1.85 | 550 | 640 | 24.6 | 52 | 17.3 | 960 | A | |
| Mg–RE–Zn–Zr | (ZRE1) RE 2.7 / ZN 2.2 / Zr 0.7 | AC T5 | 1.80 | 545 | 640 | 26.8 | 100 | 7.3 | 1050 | A | B |
| | (RZ5) Zn 4.0 / RE 1.2 / Zr 0.7 | AC T5 | 1.84 | 510 | 640 | 27.1 | 113 | 6.8 | 960 | B | |
| | (ZE63) Zn 6 / RE 2.5 / Zr 0.7 | AC T6 | 1.87 | 515 | 630 | 27.0 | 109 | 5.6 | 960 | A | |
| Mg–Th–Zn–Zr** | (ZTY) Th 0.8 / Zn 0.5 / Zr 0.6 | T1 | 1.76 | 600 | 645 | 26.4 | 121 | 6.3 | 960 | A | |
| | (ZT1) Th 3.0 / Zn 2.2 / Zr 0.7 | AC T5 | 1.83 | 550 | 647 | 26.7 | 105 | 7.2 | 960 | A | (B) |
| | (TZ6) Zn 5.5 / Th 1.8 / Zr 0.7 | AC T5 | 1.87 | 500 | 630 | 27.6 | 113 | 6.6 | 960 | B | |
| Mg–Ag–RE–Zr | (QE22) Ag 2.5 / RE(D) 2.0 / Zr 0.6 | AC T6 | 1.82 | 550 | 640 | 26.7 | 113 | 6.85 | 1000 | A | |
| | (EQ21) RE(D) 2.2 / Ag 1.5 / Cu 0.07 / Zr 0.7 | AC T6 | 1.81 | 540 | 640 | 26.6 | 113 | 6.85 | 1000 | A | |

*continued overleaf*

**Table 2.4**  *(continued)*

| Material | Nominal composition † % | | Condition | Density at 20°C g cm⁻³ | Melting point °C Sol. | Liq. | Coeff. of thermal expansion 20–200°C $10^{-6}$ K⁻¹ | Thermal conductivity W m⁻¹ K⁻¹ | Electrical resistivity μΩ cm | Specific heat 20–200°C J kg⁻¹ K⁻¹ | Weldability by argon arc process‡ | Relative damping capacity§ |
|---|---|---|---|---|---|---|---|---|---|---|---|---|
| Mg–Zn–Cu–Mn | (ZC63)Zn | 6.0 | AC T6 | 1.87 | 465 | 600 | 26.0 | 122 | 5.4 | 962 | B | |
| | Cu | 2.7 | | | | | | | | | | |
| | Mn | 0.5 | | | | | | | | | | |
| | (ZC71)Zn | 6.5 | T6 | 1.87 | 465 | 600 | 26.0 | 122 | 5.4 | 62 | B | |
| | Cu | 1.3 | | | | | | | | | | |
| | Mn | 0.8 | | | | | | | | | | |
| MG–Ag–RE–** | | | | | | | | | | | | |
| Th–Zr | (QH21)Ag | 2.5 | AC T6 | 1.82 | 540 | 640 | 26.7 | 113 | 6.85 | 1005 | A | – |
| | RE(D) | 1.0 | | | | | | | | | | |
| | Th | 1.0 | | | | | | | | | | |
| | Zr | 0.7 | | | | | | | | | | |
| Mg–Zr | (ZA)Zr | 0.6 | AC | 1.75 | 650 | 651 | 27.0 | (146) | (4.5) | 1050 | A | A |

AC Sand cast.
T4 Solution heat treated.

T5 Precipitation heat treated.
T6 Fully heat treated.

† Mg–Al type alloys normally
contain 0.2–0.4% Mn to improve
corrosion resistance.
** Thorium containing alloys are
being replaced by alternative Mg
alloys.

T1 Extruded, rolled or forged.
RE Cerium mischmetal containing
approx. 50% Ce.
* Non-equilibrium solidus 420°C.
() Estimated value.

RE(D) Mischmetal enriched in
neodynium.

RE(Δ) Neodynium + Heavy Rare
Earths.

‡ Weldability rating:
A Fully weldable.

B Weldable.

C Not recommended where fusion
welding is involved.

§ Damping capacity rating:
A Outstanding.

B Equivalent to cast iron.

C Inferior to cast iron but better than Al-base
cast alloys.

## 2.4    The physical properties of titanium and titanium alloys

Table 2.5    PHYSICAL PROPERTIES OF TITANIUM AND TITANIUM ALLOYS AT NORMAL TEMPERATURES

| Material IMI designation | Nominal composition % | | Density g cm$^{-3}$ | Coefficient of expansion 20–100°C 10$^{-6}$ K$^{-1}$ | Thermal conductivity 20–100°C W m$^{-1}$ K$^{-1}$ | Resistivity 20°C μΩ cm | Temp. coefficient of resistivity 20–100°C μΩ cm K$^{-1}$ | Specific heat 50°C J kg$^{-1}$ K$^{-1}$ | Magnetic suscept. 10$^{-6}$ cgs units g$^{-1}$ |
|---|---|---|---|---|---|---|---|---|---|
| CP Titanium | Commercially pure | | 4.51 | 7.6 | 16 | 48.2 | 0.0022 | 528 | +3.4 |
| IMI 230 | Cu | 2.5 | 4.56 | 9.0 | 13 | 70 | 0.0026 | – | – |
| IMI 260/261 | Pd | 0.2 | 4.52 | 7.6 | 16 | 48.2 | 0.0022 | 528 | – |
| IMI 315 | Al | 2.0 | 4.51 | 6.7 | 8.4 | 101.5 | 0.0003 | 460 | +4.1 |
|  | Mn | 2.0 | | | | | | | |
| IMI 317 | Al | 5.0 | 4.46 | 7.9 | 6.3 | 163 | 0.0006 | 470 | +3.2 |
|  | Sn | 2.5 | | | | | | | |
| IMI 318 | Al | 6.0 | 4.42 | 8.0 | 5.8 | 168 | 0.0004 | 610 | +3.3 |
|  | V | 4.0 | | | | | | | |
| IMI 550 | Al | 4.0 | 4.60 | 8.8 | 7.9 | 159 | 0.0004 | – | – |
|  | Mo | 4.0 | | | | | | | |
|  | Sn | 2.0 | | | | | | | |
|  | Si | 0.5 | | | | | | | |
| IMI 551 | Al | 4.0 | 4.62 | 8.4 | 5.7 | 170 | 0.0003 | 400 | +3.1 |
|  | Mo | 4.0 | | | | | | | |
|  | Sn | 4.0 | | | | | | | |
|  | Si | 0.5 | | | | | | | |
| IMI 679 | Sn | 11.0 | 4.84 | 8.0 | 7.1 | 163 | 0.0004 | – | – |
|  | Zr | 5.0 | | | | | | | |
|  | Al | 2.25 | | | | | | | |
|  | Mo | 1.0 | | | | | | | |
|  | Si | 0.2 | | | | | | | |
| IMI 680 | Sn | 11.0 | 4.86 | 8.9 | 7.5 | 165 | 0.0003 | – | – |
|  | Mo | 4.0 | | | | | | | |
|  | Al | 2.25 | | | | | | | |
|  | Si | 0.2 | | | | | | | |
| IMI 685 | Al | 6.0 | 4.45 | 9.8 | 4.8 | 167 | 0.0004 | – | – |
|  | Zr | 5.0 | | | | | | | |
|  | Mo | 0.5 | | | | | | | |
|  | Si | 0.25 | | | | | | | |
| IMI 829 | Al | 5.5 | 4.53 | 9.45 | 7.8 | – | – | 530 | – |
|  | Sn | 3.5 | | | | | | | |
|  | Zr | 3.0 | | | | | | | |
|  | Nb | 1.0 | | | | | | | |
|  | Mo | 0.3 | | | | | | | |
|  | Si | 0.3 | | | | | | | |
| IMI 834 | Al | 5.8 | 4.55 | 10.6 | – | – | – | – | – |
|  | Sn | 4.0 | | | | | | | |
|  | Zr | 3.5 | | | | | | | |
|  | Nb | 0.7 | | | | | | | |
|  | Mo | 0.5 | | | | | | | |
|  | Si | 0.35 | | | | | | | |
|  | C | 0.06 | | | | | | | |

# 3 Mechanical properties of light metals and alloys

The following tables summarize the mechanical properties of the more important industrial light metals and alloys.

In the tables of tensile properties at normal temperature the nominal composition of the alloys is given, followed by the appropriate British and other specification numbers. Most specifications permit considerable latitude in both composition and properties, but the data given in these tables represent typical average values which would be expected from materials of the nominal composition quoted, unless otherwise stated. For design purposes it is essential to consult the appropriate specifications to obtain minimum and maximum values and special conditions where these apply.

The data in the tables referring to properties at elevated and at sub-normal temperatures, and for creep, fatigue and impact strength have been obtained from a more limited number of tests and sometimes from a single example. In these cases the data refer to the particular specimens tested and cannot be relied upon as so generally applicable to other samples of material of the same nominal composition.

## 3.1 Mechanical properties of aluminium and aluminium alloys

The compositional specifications for wrought aluminium alloys are now internationally agreed throughout Europe, Australia, Japan and the USA. The system involves a four-digit description of the alloy and is now specified in the UK as BS EN 573, 1995. Registration of wrought alloys is administered by the Aluminum Association in Washington, DC. International agreement on temper designations has been achieved, and the standards agreed for the European Union, the Euro-Norms, are replacing the former British Standards. Thus BS EN 515. 1995 specifies in more detail the temper designations to be used for wrought alloys in the UK. At present, there is no Euro-Norm for cast alloys and the old temper designations are still used for cast alloys.

In the following tables the four-digit system is used, wherever possible, for wrought materials.

### 3.1.1 Alloy designation system for wrought aluminium

The first of the four digits in the designation indicates the alloy group according to the major alloying elements, as follow:

| | |
|---|---|
| 1XXX | aluminium of 99.0% minimum purity and higher |
| 2XXX | copper |
| 3XXX | manganese |
| 4XXX | silicon |
| 5XXX | magnesium |
| 6XXX | magnesium and silicon |
| 7XXX | zinc |
| 8XXX | other element, incl. lithium |
| 9XXX | unused |

1XXX Group:
: In this group the last two digits indicate the minimum aluminium percentage. Thus 1099 indicates aluminium with a minimum purity of 99.99%. The second digit indicates modifications in impurity or alloying element limits. 0 signifies unalloyed aluminium and integers 1 to 9 are allocated to specific additions.

2XXX-8XXX Groups:
: In these groups the last two digits are simply used to identify the different alloys in the groups and have no special significance. The second digit indicates alloy modifications, zero being allotted to the original alloy.

National variations of existing compositions are indicated by a letter after the numerical designation, allotted in alphabetical sequence, starting with A for the first national variation registered. The specifications and properties for Cast Aluminium Alloys are tabulated in Chapter 4.

### 3.1.2　Temper designation system for aluminium alloys

The following tables use the internationally agreed temper designations for wrought alloys, (BS EN 515. 1995) and the more frequently used ones are listed below. The old ones still used for existing BS specifications e.g. BS 1490. 1989 for castings are compared with the new ones at the end of this section.

| U.K. | Meaning |
|---|---|
| F | As manufactured or fabricated |
| H111 | Fully soft annealed condition |
| Strain-hardened alloys | |
| H | Strain hardened non-heat-treatable material |
| H1x | Strain hardened only |
| H2x | Strain hardened only and partially annealed to achieve required temper |
| H3x | Strain hardened only and stabilized by low temperature heat treatment to achieve required temper |
| H12,H22,H32 | Quarter hard, equivalent to about 20–25% cold reduction |
| H14,H24,H34 | Half hard, equivalent to about 35% cold reduction |
| H16,H26,H36 | Three-quarter hard, equivalent to 50–55% cold reduction |
| H18,H28,H38 | Fully hard, equivalent to about 75% cold reduction |
| Heat-treatable alloys | |
| T1 | Cooled from an Elevated Temperature Shaping Process and aged naturally to a substantially stable condition |
| T2 | Cooled from an Elevated Temperature Shaping Process, cold worked and aged naturally to a substantially stable condition |
| T3 | Solution heat-treated, cold worked and aged naturally to a substantially stable condition |
| T4 | Solution heat-treated and aged naturally to a substantially stable condition |
| T5 | Cooled from an Elevated Temperature Shaping Process and then artificially aged |
| T6 | Solution heat-treated and then artificially aged |
| T7 | Solution heat-treated and then stabilized (over-aged) |
| T8 | Solution heat-treated, cold worked and then artificially aged |
| T9 | Solution heat-treated, artificially aged and then cold worked |
| T10 | Cooled from an Elevated Temperature Shaping Process, artificially aged and then cold worked |

A large number of variants in these tempers has been introduced by adding additional digits to the above designations. For example, the addition of the digit 5 after T1-9 signifies that a stress relieving treatment by stretching has been applied after solution heat-treatment.

A full list is given in BS EN 515. 1995 but some of the more common ones used in the following tables are given below.

| | |
|---|---|
| T351 | Solution heat-treated, stress-relieved by stretching a controlled amount (usually 1–3% permanent set) and then naturally aged. There is no further straightening after stretching. This applies to sheet, plate, rolled rod and bar and ring forging. |
| T3510 | The same as T351 but applied to extruded rod, bar, shapes and tubes. |
| T3511 | As T3510, except that minor straightening is allowed to meet tolerances. |
| T352 | Solution heat-treated, stress-relieved by compressing (1–5% permanent set) and then naturally aged. |
| T651 | Solution heat-treated, stress-relieved by stretching a controlled amount (usually 1–3% permanent set) and then artificially aged. There is no further straightening after stretching. This applies to sheet, plate, rolled rod and bar and ring forging. |
| T6510 | The same as T651 but applied to extruded rod, bar, shapes and tubes. |
| T6511 | As T6510, except that minor straightening is allowed to meet tolerances. |
| T73 | Solution heat-treated and then artificially overaged to improve corrosion resistance. |
| T7651 | Solution heat-treated, stress-relieved by stretching a controlled amount (Again about 1–3% permanent set) and then artificially over-aged in order to obtain a good resistance to exfoliation corrosion. There is no further straightening after stretching. This applies to sheet, plate, rolled rod and bar and to ring forging. |
| T76510 | As T7651 but applied to extruded rod, bar, shapes and tubes. |
| T76511 | As T7510, except that minor straightening is allowed to meet tolerances. |

In some specifications, the old system is still being applied. The equivalents between old and new are as follows.

| BS EN 515 | BS1470/90 | Pre-1969 BS |
|---|---|---|
| | F | M |
| H111 | 0 | 0 |
| T3 | TD | WD |
| T4 | TB | W |
| T5 | TE | P |
| T6 | TF | WP |
| T8 | TH | WDP |

TH7 is as TH and then stabilised.

F/M is as manufactured or fabricated.

**Table 3.2** ALUMINIUM AND ALUMINIUM ALLOYS-MECHANICAL PROPERTIES AT ROOM TEMPERATURE

*Wrought Alloys*

| Specification | Nominal composition % | Form | Condition | 0.2% Proof stress MPa | Tensile strength MPa | Elong. % on 50 mm ($\geq$2.6 mm) or $5.65\sqrt{S_0}$ | Shear strength MPa | Brinell hardness ($P = 5D^2$) | Fatigue strength (unnotched) 500 MHz MPa | Impact energy J | Fracture toughness ($MPa\,m^{1/2}$) | Remarks |
|---|---|---|---|---|---|---|---|---|---|---|---|---|
| 1199 | Al 99.99 | Sheet | H111 | 20 | 55 | 55 | 50 | 15 | – | – | – | Highest quality reflectors |
|  |  |  | H14 | 60 | 85 | 20 | 60 | 23 | – | – | – |  |
|  |  |  | H18 | 85 | 110 | 12 | 70 | 28 | – | – | – |  |
| 1080A | Al 99.8 | Sheet | H111 | 25 | 70 | 50 | 60 | 19 | – | – | – | Domestic trim, chemical plant |
|  |  |  | H14 | 95 | 100 | 17 | 70 | 29 | – | – | – |  |
|  |  |  | H18 | 125 | 135 | 11 | 70 | 29 | – | – | – |  |
|  |  | Wire | H111 | – | 70 | – | 60 | 19 | – | – | – |  |
|  |  |  | H14 | 90 | 105 | – | 70 | 30 | – | – | – |  |
|  |  |  | H18 | 110–140 | 130–160 | – | – | 35–41 | – | – | – |  |
| 1050A | Al 99.5 | Sheet | H111 | 35 | 80 | 47 | 65 | 21 | – | – | – | General purpose formable alloy |
|  |  |  | H14 | 105 | 110 | 15 | 75 | 30 | – | – | – |  |
|  |  |  | H18 | 130 | 145 | 10 | 85 | 40 | – | – | – |  |
|  |  | Bars and sections as extruded | H15 | 50 | 75 | 38 | 65 | 22 | – | – | – |  |
|  |  | Rivet stock | H111 | 125 | 140 | – | – | – | – | – | – |  |
|  |  | Tubes | H111 | – | 75 | – | 65 | 21 | – | – | – |  |
|  |  |  | H18 < 75 mm | 120 | 125 | – | 75 | – | – | – | – |  |
|  |  |  | H18 > 75 mm | 110 | 115 | – | 70 | – | – | – | – |  |
|  |  | Wire | H111 | 42 | 75 | – | 65 | 21 | – | – | – |  |
|  |  |  | H14 | 100 | 115 | – | 75 | 30 | – | – | – |  |
|  |  |  | H18 | 115–170 | 140–195 | – | – | 38–48 | – | – | – |  |
| 1350 | Al 99.5 | Wire | H111 | 28 | 83 | – | 55 | – | – | – | – | Electrical conductors |
|  |  |  | H14 | 97 | 110 | – | 69 | – | – | – | – |  |
|  |  |  | H18 | 165 | 186 | – | 103 | – | 48 | – | – |  |

*continued overleaf*

**Table 3.2** (continued)

Wrought Alloys

| Specification | Nominal composition % | Form | Condition | 0.2% Proof stress MPa | Tensile strength MPa | Elong. % on 50 mm (≥2.6 mm) or 5.65√$S_0$ | Shear strength MPa | Brinell hardness ($P = 5D^2$) | Fatigue strength (unnotched) 500 MHz MPa | Impact energy J | Fracture toughness (MPa·m$^{1/2}$) | Remarks |
|---|---|---|---|---|---|---|---|---|---|---|---|---|
| 1200 | Al 99.0 | Sheet | H111 | 35 | 90 | 43 | 70 | 22 | 35 | 27 | – | General purpose, slightly higher strength than 105A |
| | | | H13 | 95 | 105 | 20 | 75 | 31 | 40 | – | – | |
| | | | H14 | 115 | 120 | 12 | 80 | 35 | 50 | 31 | – | |
| | | | H16 | 125 | 135 | 11 | 90 | 38 | 60 | – | – | |
| | | | H18 | 145 | 160 | 9 | 95 | 42 | 60 | 26 | – | |
| | | Bars and sections as extruded | H111 | 40 | 85 | 38 | 70 | 23 | 45* | 27 | – | |
| | | Tubes | H111 | – | 90 | 40 | 70 | 21 | – | – | – | |
| | | | H > 75 mm | 128 | 131 | 6 | 100 | 34 | – | – | – | |
| | | | H < 75 mm | 120 | 124 | 6 | 95 | 32 | – | – | – | |
| 2011 | Cu 5.5 Bi 0.5 Pb 0.5 | Extruded bar | T3 25 mm | 295 | 340 | 14 | 240 | 95 | – | – | – | Free machining alloy |
| | | | T6 50–75 mm | 260 | 370 | 16 | 240 | 100 | – | – | – | |
| | | Wire | T3 ≤ 10 mm | 350 | 365 | – | – | – | – | – | – | |
| 2014 | Cu 4.4 Mg 0.7 Si 0.8 Mn 0.75 | Plate | T451 | 290 | 425 | 22 | 260 | 108 | 140 | – | – | Heavy duty applications in transport and aerospace, e.g. large parts, wings |
| | | | T651 | 415 | 485 | 10 | 290 | 139 | 125 | – | – | |
| | | Bar/tube | T6510 | 440 | 490 | 8 | – | – | – | – | – | |
| 2014A | Cu 4.4 Mg 0.7 Si 0.8 Mn 0.75 | Sheet | T4 | 270 | 450 | 20 | 260 | 115 | 130* | – | – | Aircraft applications (cladding when used 1070A) |
| | | | T6 | 430 | 480 | 10 | 295 | 135 | 130* | – | – | |
| | | Clad sheet | T4 | 250 | 425 | 22 | 250 | – | 95* | – | – | |
| | | | T6 | 385 | 440 | 10 | 260 | – | 95* | – | – | |
| | | Bars and sections | T4 | 315 | 465 | 17 | – | 115 | 140 | 22 | – | |
| | | | T6 | 465 | 500 | 10 | – | 135 | 124 | 8 | – | |
| | | Tubes | T4 | 310 | 425 | 12 | – | 115 | – | – | – | |
| | | | T6 | 415 | 480 | 9 | – | 135 | – | – | – | |
| | | Wire | T4 | 340 | 445 | 15 | – | 115 | – | – | – | |
| | | | T6 | 425 | 465 | – | – | 135 | – | – | – | |
| | | River stock | T4 | 340 | 450 | – | – | – | – | – | – | |
| | | Bolt and screw stock | T6 | 425 | 460 | – | – | – | – | – | – | |
| 2024 | Cu 4.5 Mg 1.5 Mn 0.6 | Plate | T3 | 345 | 485 | 18 | 285 | 120 | 140 | – | – | Structural applications, especially transport and aerospace |
| | | | T351 | 325 | 470 | 19 | 285 | 120 | 140 | – | – | |

| Alloy | Composition | Form | Temper | | | | | | | | | Applications |
|---|---|---|---|---|---|---|---|---|---|---|---|---|
| 2024 | Cu 4.5 Mg 1.5 Mn 0.6 | Plate/sheet extrusions | H111 | 75 | 185 | 20 | 125 | 47 | 90 | – | – | Aircraft structures |
| | | | T4 | 325 | 470 | 20 | 285 | 120 | 140 | – | – | |
| | | | T6 | 395 | 475 | 10 | – | – | – | – | – | |
| 2117 | Cu 2.5 Si 0.6 Mg 0.4 | Sheet | T4 | 165 | 295 | 24 | 195 | 70 | 95 | – | – | Vehicle body sheet |
| 2090 | Cu 2.7 Li 2.7 Zr 0.12 | Plate | T81 | 517 | 550 | 8 | – | – | – | – | 71 | High strength, low density aero-alloy |
| | | Plate (12.5 mm) | T81 | 535 | 565 | 11 | – | – | – | – | 34 | |
| 2091 | Cu 2.1 Li 2.0 Mg 1.50 Zr 0.1 | Plate (12 mm) | T8 × 51 | 310 | 420 | 14 | – | – | – | – | – | Medium strength, low density aero-alloy in damage-tolerant temper |
| | | Plate (40 mm) | T8 × 51 | 310 | 430 | 6 | – | – | – | – | – | |
| | | Extrusion (10 mm) | T851 | 505 | 580 | 7 | – | – | – | – | – | |
| | | Extrusion (30 mm) | T851 | 465 | 520 | 11 | – | – | – | – | 35 | |
| | | Plate (12 mm) | T851 | 460 | 525 | 10 | – | – | – | – | 43 | Medium strength, low density aero-alloy |
| | | Plate (38 mm) | T851 | 430 | 495 | 8 | – | – | – | – | 38 | Medium strength, low density aero-alloy |
| | | Sheet | T8 | 390 | 495 | 10 | – | – | – | – | 38 | |
| 2219 | Cu 6 Mn 0.3 V 0.1 | Plate/sheet/ forgings | H111 | 75 | 170 | 18 | – | – | – | – | – | Weldable, creep resistant, high-temperature aerospace applications |
| | | | T4 | 185 | 360 | 20 | – | – | – | – | – | |
| | | | T6 | 290 | 415 | 10 | – | – | 105 | – | – | |
| 2004 | Cu 6 Zr 0.4 | Sheet | H111 | 150 | 230 | 15 | – | – | 100* | – | – | Superplastically deformable sheet |
| | | | T6 | 300 | 420 | 12 | – | – | 150* | – | – | |
| 2031 | Cu 2.3 Ni 1.0 Mg 0.9 Si 0.9 Fe 0.9 | Forgings | T4 | 235 | 355 | 22 | 201 | 95 | – | – | – | Aero-engines, missile fins |
| | | | T6 | 340 | 420 | 15 | 201 | 95 | – | – | – | |
| 2618A | Cu 2.0 Mg 1.5 Si 0.9 Fe 0.9 Ni 1.0 | Forgings | H111 | 70 | 170 | 20 | – | 45 | 85* | – | – | Aircraft engines |
| | | | T6 | 330 | 430 | 8 | 295 | 130 | 170* | – | – | |

*continued overleaf*

**Table 3.2** *(continued)*

*Wrought Alloys*

| Specification | Nominal composition % | Form | Condition | 0.2% Proof stress MPa | Tensile strength Mpa | Elong.% on 50 mm (≥2.6 mm) or 5.65√S₀ | Shear strength MPa | Brinell hardness (P = 5D²) | Fatigue strength (unnotched) 500 MHz MPa | Impacy energy J | Fracture toughness (MPa m^{1/2}) | Remarks |
|---|---|---|---|---|---|---|---|---|---|---|---|---|
| 3103 | Mn 1.25 | Sheet | H111 | 65 | 110 | 40 | 80 | 30 | 50 | 34 | – | General purpose, holloware, building sheet |
| | | | H12 | 125 | 130 | 17 | 90 | 40 | 55 | – | – | |
| | | | H14 | 140 | 155 | 11 | 95 | 44 | 60 | 29 | – | |
| | | | H16 | 160 | 180 | 8 | 105 | 47 | 70 | – | – | |
| | | | H18 | 185 | 200 | 7 | 110 | 51 | 70 | 20 | – | |
| | | Wire | H111 | 60 | 115 | – | – | 30 | – | – | – | |
| | | | H14 | 135 | 155 | – | – | 45 | – | – | – | |
| | | | H18 | 170–200 | 205–245 | – | – | 55–65 | – | – | – | Building cladding sheet |
| 3105 | Mn 0.35 Mg 0.6 | Sheet | H111 | 55 | 115 | 24 | 85 | – | – | – | – | |
| | | | H14 | 150 | 170 | 5 | 105 | – | – | – | – | |
| | | | H18 | 195 | 215 | 3 | 115 | – | – | – | – | |
| 3004 | Mn 1.2 Mg 1.0 | Sheet | H111 | 70 | 180 | 20 | 110 | 45 | 95 | – | – | Sheet metal work, storage tanks |
| | | | H14 | 200 | 240 | 9 | 125 | 63 | 105 | – | – | |
| | | | H18 | 250 | 285 | 5 | 145 | 77 | 110 | – | – | |
| 3008 | Mn 1.6 Fe 0.7 Zr 0.3 | Sheet | H111 | 50 | 120 | 23 | – | – | – | – | – | Thermally reistant alloy. Vitreous enamelling |
| | | | H18 | 270 | 280 | 4 | – | – | – | – | – | |
| 3003 clad with 4343 | Mn 1.2 Si 7.5 | Sheet | H111 | 40 | 110 | 30 | 75 | – | – | – | – | Flux brazing sheet |
| | | | H12 | 125 | 130 | 10 | 85 | – | – | – | – | |
| | | | H14 | 145 | 150 | 8 | 95 | – | – | – | – | |
| | | | H16 | 170 | 175 | 5 | 105 | – | – | – | – | |
| 3003 clad with 4004 | Mn 1.2 Si 1.0 Mg 1.5 | Sheet | Physical properties | as for 3003 clad with 4343 | | | | | | | | Vacuum brazing sheet |

| Alloy | Composition | Form | Temper | | | | | | | | | Applications |
|---|---|---|---|---|---|---|---|---|---|---|---|---|
| 4032 | Si 12.0 Cu 1.0 Mg 1.0 Ni 1.0 | Forgings | T6 | 240 | 325 | 5 | – | 115 | 110 | – | – | Pistons |
| 4043A | Si 12.0 | Rolled wire | | 75 | 130 | 20 | – | – | – | – | – | Welding filler wire |
| 4047A | Si 5.0 | Wire | F | 189 | 225 | 8 | – | – | – | – | – | Brazing rod |
| 5657 | Mg 0.8 | Sheet | H111 | 40 | 110 | 25 | 75 | 28 | – | – | – | High base purity, bright trim alloy |
| | | | H14 | 140 | 160 | 12 | 95 | 40 | – | – | – | |
| | | | H18 | 165 | 195 | 7 | 105 | 50 | – | – | – | |
| 5005 | Mg 0.8 | Sheet | H111 | 40 | 125 | 25 | 75 | 28 | – | – | – | Architectural trim, commercial vehicle trim |
| | | | H14 | 150 | 160 | 6 | 95 | – | – | – | – | |
| | | | H18 | 195 | 200 | 4 | 110 | – | – | – | – | |
| 5251 | Mg 2.25 Mn 0.25 | Sheet | H111 | 95 | 185 | 22 | 125 | 45 | 110 | 50 | – | Sheet metal work |
| | | | H14 | 230 | 245 | 7 | 145 | 70 | 125 | 29 | – | |
| | | | H18 | 275 | 285 | 2 | 175 | 80 | 140 | – | – | |
| 5251 | Mg 2.0 Mn 0.3 | Bar | F | 60 | 170 | 16 | – | – | – | – | – | Marine and transport applications; good workability combined with good corrosion resistance and high fatigue resistance |
| | | Sheet | H111 | 60 | 180 | 20 | 125 | 47 | – | – | – | |
| | | | H22 | 130 | 220 | 8 | 132 | 65 | 92 | – | – | |
| | | | H24 | 175 | 250 | 5 | 139 | 74 | 124 | – | – | |
| | | | H28 | 215 | 270 | 4 | – | – | – | – | – | |
| | | Bars and sections as extruded (F) | | 95 | 185 | 20 | 125 | 45 | 95* | 49 | – | |
| | | Tubes | H111 | 100 | 200 | 20 | – | – | – | – | – | |
| | | | H14 | 230 | 250 | 6 | – | – | – | – | – | |
| | | | H18 | 255 | 270 | 5 | – | – | – | – | – | |
| | | Wire | H111 | 95 | 200 | – | – | 48 | – | – | – | |
| | | | H18 | 260–290 | 280–310 | – | – | 75–85 | – | – | – | |

*continued overleaf*

**Table 3.2** (continued)

Wrought Alloys

| Specification | Nominal composition % | Form | Condition | 0.2% Proof stress MPa | Tensile strength MPa | Elong.% on 50 mm (≥2.6 mm) or 5.65√S₀ | Shear strength MPa | Brinell hardness (P = 5D²) | Fatigue strength (unnotched) 500 MHz MPa | Impact energy J | Fracture toughness (MPa m^{1/2}) | Remarks |
|---|---|---|---|---|---|---|---|---|---|---|---|---|
| 5154A | Mg 3.5 Mn 0.5 | Sheet | H111 | 125 | 240 | 24 | 155 | 55 | 115 | – | – | Welded structures, storage tanks, salt water service |
| | | | H22 | 245 | 295 | 10 | 175 | 80 | 125 | – | – | |
| | | | H24 | 275 | 310 | 9 | 175 | 95 | 130 | – | – | |
| | | Bars and sections as extruded (F) | | 125 | 230 | 25 | 145 | 55 | 140* | 48 | – | |
| | | Tubes | H111 | 125 | 225 | 20 | – | 55 | – | – | – | |
| | | | H14 | 220 | 280 | 7 | – | – | – | – | – | |
| | | Wire | H111 | 125 | 240 | – | – | 55 | – | – | – | |
| | | | H14 | 265 | 295 | – | – | 90 | – | – | – | |
| | | | H18 | 310 | 355 | – | – | 100 | – | – | – | |
| | | Rivet stock | H111 | 125 | 250 | – | – | – | – | – | – | |
| | | | H12 | – | 290 | – | – | – | – | – | – | |
| 5454 | Mg 2.7 Mn 0.75 Cr 0.12 | Sheet | H111 | 105 | 250 | 22 | 159 | 65 | 115 | – | – | Higher strength alloy for marine and transport, pressure vessels and welded structures |
| | | | H22 | 200 | 277 | 7 | 165 | 77 | 125 | – | – | |
| | | | H24 | 225 | 297 | 5 | 179 | 85 | 130 | – | – | |
| 5083 | Mg 4.5 Mn 0.7 Cr 0.15 | Sheet | H111 | 170 | 310 | 21 | 170 | 72 | – | – | – | Marine applications, cryogenics, welded pressure vessels. |
| | | | H24 | 290 | 370 | 9 | 210 | 110 | – | – | – | |
| | | Bars and sections as extruded (F) | | 180 | 315 | 19 | 180 | 77 | – | – | – | |
| 5083 | Mg 4.5 Mn 0.7 Cr 0.15 | Tube | H111 | 180 | 320 | 20 | – | 77 | – | – | – | |
| | | | H14 | 300 | 375 | 7 | – | – | – | – | – | |
| 5556A | Mg 5 | Wire | H14 | 250 | 330 | 12 | – | – | – | – | – | Weld filler wire |
| 5056A | Mg 5.0 Mn 0.5 | Wire | H111 | 140 | 300 | – | – | 65 | – | – | – | Rivets, bolts, screws |
| | | | H14 | 300 | 340 | – | – | 95 | – | – | – | |
| | | | H18 | 340–400 | 400–450 | – | – | 110–120 | – | – | – | |
| | | Rivet stock | H111 | 140 | 300 | – | – | 65 | – | – | – | |
| | | | H12 | – | 350 | – | – | – | – | – | – | |
| | | Bolt and screw stock | H14 | 300 | 340 | – | – | – | – | – | – | |

| Alloy | Composition | Form | Temper | | | | | | | | | Applications |
|---|---|---|---|---|---|---|---|---|---|---|---|---|
| 6060 | Mg 0.5 Si 0.4 | Bar | T4 | 90 | 150 | 20 | – | – | – | – | – | Medium strength extrusion alloy for doors, windows, pipes, architectural use; weldable and corrosion-resistant |
| | | | T5 | 130 | 175 | 13 | – | – | – | – | – | |
| | | | T6 | 190 | 220 | 13 | – | – | – | – | – | |
| 6063 | Mg 0.5 Si 0.5 | Bars, sections and forgings | F | 85 | 155 | 30 | 100 | 35 | – | – | – | Architectural extrusions (fast extruding) |
| | | | T4 | 115 | 180 | 30 | 130 | 52 | 60 | 43 | – | |
| | | | T6 | 210 | 245 | 20 | 160 | 75 | 70 | 31 | – | |
| | | Wire | H111 | – | 115 | – | – | – | – | – | – | |
| | | | T4 | 115 | 180 | – | – | 50 | – | – | – | |
| | | | T6 | 195 | 230 | – | – | 70 | – | – | – | |
| 6063A | Mg 1.0 Si 0.5 | Bar | T6510 | 280 | 310 | 22 | 129 | 50 | 79 | – | – | Transport, windows, furniture, doors and architectural uses, pipes (irrigation) |
| | | | T5 | 160 | 200 | 12 | 117 | 65 | 69 | – | – | |
| | | | T6 | 210 | 240 | 12 | 152 | 78 | 85 | – | – | |
| 6061 | Mg 1.0 Si 0.6 Cr 0.25 Cu 0.2 | Bars and sections | T4 | 145 | 230 | 20 | 160 | 60 | – | 34 | – | Intermediate strength extrusion alloy |
| | | | T6 | 280 | 310 | 13 | 200 | 90 | – | 27 | – | |
| | | Wire | T8 ≤ 6 mm | 310–400 | 385–430 | – | – | – | – | – | – | |
| | | | T8 (6–10 mm) | 295–385 | 380–415 | – | – | – | – | – | – | |
| | | Bar | T6510 | 280 | 310 | 13 | 205 | 100 | 95 | – | – | |
| | | Bolt and screw stock | T8 | 290 | 340 | – | – | – | – | – | – | |
| 6082 | Mg 1.0 Si 1.0 Mn 0.7 | Bar/extrusion | T5 | 260 | 300 | 15 | 185 | 85 | – | – | – | |
| | | | T6510 | 285 | 315 | 11 | – | – | – | – | – | |
| | | Plate | T451 | 150 | 240 | 19 | – | 68 | – | – | – | |
| | | | T651 | 289 | 315 | 12 | 205 | 104 | – | – | – | |
| | | | T6 | 285 | 315 | 12 | – | 100 | – | – | – | |
| | | | T4 | 160 | 240 | 25 | 180 | 65 | – | 41 | – | |
| | Mg 1.0 | Bars, sections and forgings | T6 | 285 | 310 | 13 | 215 | 100 | – | 34 | – | |
| | Si 1.0 Mn 0.5 | Tubes | T4 | 160 | 245 | 20 | – | 65 | – | – | – | |
| | | | T6 | 285 | 325 | 10 | – | 95 | – | – | – | |

*continued overleaf*

**Table 3.2** (continued)

Wrought Alloys

| Specification | Nominal composition % | Form | Condition | 0.2% Proof stress MPa | Tensile strength Mpa | Elong.% on 50 mm (≥2.6 mm) or $5.65\sqrt{S_0}$ | Shear strength MPa | Brinell hardness ($P = 5D^2$) | Fatigue strength (unnotched) 500 MHz MPa | Impact energy J | Fracture toughness (MPa m$^{1/2}$) | Remarks |
|---|---|---|---|---|---|---|---|---|---|---|---|---|
| 6463 | Mg 0.55 Si 0.4 | Bar | T4 T6 | 130 215 | 180 240 | 16 12 | – 150 | 55 79 | 70 70 | – – | – – | Vehicle body sheet |
| 6009 | Si 0.8 Mg 0.6 Mn 0.5 Cu 0.4 | Sheet | T4 T6 | 130 325 | 235 345 | 24 12 | 205 150 | 60 – | 97 115 | – – | – – | |
| 7020 | Zn 4.5 Mg 1.2 Zr 0.15 | Bars and sections | T4 T6 | 225 310 | 340 370 | 18 15 | – – | 100 126 | – – | – – | – – | Transportable bridging |
| 7075 | Zn 5.6 Mg 2.5 Cu 1.6 Cr 0.25 | Sheet/plate/ forgings/ extrusion | H111 T4 T73 | 105 505 435 | 230 570 505 | 17 11 13 | 150 330 – | 60 150 – | – 160 – | – 7 – | – – – | Aircraft structures |
| 7050 | Zn 6.2 Mg 2.2 Cu 2.3 Zr 0.12 | Thick section plate/ forgings | T736 | 455 | 515 | 11 | – | – | 220 | – | – | Low quench sensitivity, high stress corrosion resistance. Aircraft structures |
| 7475 | Zn 5.7 Mg 2.2 Cu 1.5 Cr 0.2 | Sheet/plate/ forgings | T61 T7351 | 525 – | 460 – | 12 – | – 270 | – – | – 220 | – – | – – | High base purity. High fracture toughness. Aircraft structures |
| 7016 | Zn 4.5 Mg 1.1 Cu 0.75 | Extrusions | T6 | 315 | 360 | 12 | – | – | – | – | – | Bright anodized vehicle bumpers |
| 7021 | Zn 5.5 Mg 1.5 Cu 0.25 Zr 0.12 | Extrusion | H111 T6 | 115 395 | 235 435 | 16 13 | – – | – – | – – | – – | – – | Bumper backing bars |

| Alloy | Composition | Form | Temper | | | | | | | | | Comments |
|---|---|---|---|---|---|---|---|---|---|---|---|---|
| 8079 | Fe 0.7 | Foil | H111 | 35 | 95 | 26 | – | – | – | – | – | Domestic foil |
| | | | H18 | 160 | 175 | 2 | – | – | – | – | – | |
| 8090 | Li 2.5 Cu 1.3 Mg 0.95 Zr 0.1 | Plate | T8151 | 483 | 518 | 4.3 | – | – | – | – | 42 | As 2090 but lower density |
| | | Plate (38/65 mm) | T8151 | 387 | 476 | 6.5 | – | – | – | – | 42.5 | Under-aged, damage-tolerant condition |
| | | | T8771 | 483 | 518 | 4.3 | – | – | – | – | 42 | Peak-aged, medium strength condition |
| | | Sheet | T81 | 360 | 420 | 11 | – | – | – | – | 42 | Under-aged, damage-tolerant condition with a recrystallized grain structure |
| | | | T6 | 373 | 472 | 5.7 | – | – | – | – | – | |
| | | | T8 | 436 | 503 | 5 | – | – | – | – | 75 | Peak-aged, medium strength condition |
| | | Extrusion | T81551 | 440 | 510 | 4 | – | – | – | – | 45 | Damage-tolerant condition |
| | | Extrusion (10 mm) | T82551 | 460 | 515 | 4.2 | – | – | – | – | 39 | Peak-aged, medium strength condition |
| | | | T851 | 515 | 580 | 5 | – | – | – | – | 30 | |
| | | Extrusion (30 mm) | T851 | 460 | 520 | 9 | – | – | – | – | 40 | |
| | | Plate (12 mm) | T851 | 455 | 500 | 7 | – | – | – | – | 33 | |
| 8090 | Li 2.4 Cu 1.2 Mg 0.50 Zr 0.14 | Forging | T651 | 468 | 517 | 7 | – | – | – | – | 28.1 | Peak-aged (32 h at 170°C. (Shrimpton) |
| | | | T651 | 400 | 453 | 7 | – | – | – | – | 36.7 | Under-aged (20 h at 150°C). (Shrimpton) |
| 8090 | Li 2.5 Cu 1.2 Mg 0.66 Zr 0.12 | Forging | | 420 | 499 | 7.8 | – | – | – | – | 16.98 | Soln. trt, 530°C, WQ, aged 30 h at 170°C |
| 8091 | Li 2.4 Cu 1.9 Mg 0.85 Zr 0.1 | | – | – | – | – | – | – | – | – | – | |

*continued overleaf*

**Table 3.2**  *(continued)*

*Wrought Alloys*

| Specification | Nominal composition % | Form | Condition | 0.2% Proof stress MPa | Tensile strength MPa | Elong.% on 50 mm (≥2.6 mm) or $5.65\sqrt{S_0}$ | Shear strength MPa | Brinell hardness ($P = 5D^2$) | Fatigue strength (unnotched) 500 MHz MPa | Fatigue Impacy energy J | Fracture toughness (MPa m$^{1/2}$) | Remarks |
|---|---|---|---|---|---|---|---|---|---|---|---|---|
| 8091 | Li 2.4 Cu 2.0 Mg 0.70 Zr 0.08 | Plate (40 mm) | | 460 | — | — | — | 164 | — | — | — | Peak-aged (6% stretch, 32 h at 170°C) |
| | | | | 408 | — | — | — | 159 | — | — | — | Peak aged (no stretch, 100 h at 170°C) |
| | | | | 408 | — | — | — | 158 | — | — | — | Duplex-aged (ditto, 24 h at RT, 48 h at 170°C) |
| 8091 | Li 2.3 Cu 1.7 Mg 0.64 Zr 0.13 | Forging | — | 436 | 503 | 8.2 | — | — | — | — | 20.72 | soln. trt. 530°C, WQ, aged 20 h at 170°C |
| | | | | | | *Cast alloys* | | | | | | |
| Al | (LMO) Al 99.0 | Sand cast | F | 30 | 80 | 30 | 55 | 25 | 30* | 19 | — | High conductivity, high ductility |
| | | Chill cast | F | 30 | 80 | 40 | 55 | 25 | 30* | 19 | — | |
| Al–Mg | (LM5) Mg 5.0 Mn 0.5 | Sand cast | F | 100 | 160 | 6 | — | 60 | 45 | 8 | — | Very high corrosion resistance |
| | | Chill cast | F | 100 | 215 | 10 | — | 65 | 95 | 12 | — | |
| | (LM10) Mg 10.0 | Sand cast | T4 | 180 | 295 | 12 | 230 | 85 | 55 | 15 | — | Strength + corrosion resistance |
| | | Chill cast | T4 | 190 | 340 | 18 | 230 | 95 | - | - | — | |
| Al–Si | (LM18) Si 5.0 | Sand cast | F | 60 | 125 | 5 | 90 | 40 | 55 | 1.5 | — | Intricate castings |
| | | Chill cast | F | 70 | 155 | 6 | 120 | 50 | 85 | 2.5 | — | |
| | (LM6)(LM20)(LM6) Si 11.5 | Sand cast | F | 65 | 170 | 8 | 110 | 55 | 45* | 4 | — | Very similar alloys, excellent casting characteristics and corrosion resistance. LM6 has slightly supperior corrosion resistance |
| | (LM6) | Chill cast | F | 75 | 215 | 10 | 130 | 60 | 60* | 9.5 | — | |
| Al–Si–Mg | (2L99) Si 7 Mg 0.4 | Sand cast | T6 | 195 | 240 | 3 | - | - | 56 | - | — | Good strength in fairly difficult castings. Cast vehicle wheels |
| | | Chill cast | T6 | 210 | 290 | 6 | - | 90 | 90 | - | — | |

| | | | | | | | | | | | | | |
|---|---|---|---|---|---|---|---|---|---|---|---|---|---|
| Al–Cu–Si | (L154) | Cu 4.2 | Sand cast | T4 | 170 | 225 | 8 | – | – | – | – | – | Aircraft castings |
| | | Si 1.2 | Chill cast | T4 | 175 | 280 | 15 | – | – | – | – | – | |
| | (L155) | | Sand cast | T6 | 215 | 295 | 5 | – | 85 | – | – | – | |
| | | | Chill cast | T6 | 215 | 320 | 10 | – | 90 | – | – | – | |
| Al–Cu–Si | (LM24) | Cu 3.5 | Chill cast | F | 110 | 200 | 3 | – | 85 | – | – | – | Excellent die casting |
| | | Si 8.0 | Die cast | F | 150 | 320 | 2 | – | 85 | – | – | – | alloy |
| | (LM4) | Cu 3.0 | Sand cast | F | 90 | 155 | 3 | – | 70 | 70* | 0.7 | – | General engineering, |
| | | Si 5.0 | Chill cast | F | 100 | 170 | 3 | – | 80 | 75* | 0.7 | – | particularly sand and |
| | | | Sand cast | T6 | 230 | 260 | 1 | – | 105 | – | – | – | permanent mould |
| | | | Chill cast | T6 | 260 | 330 | 3 | – | 110 | – | – | – | castings |
| | (LM22) | Cu 3.0<br>Si 5.0<br>Mn 0.5 | Chill cast | T4 | 115 | 260 | 9 | – | 75 | – | 4.5 | – | Good combination of impact resistance and strength |
| | (LM2) | Cu 1.5 | Sand cast | F | 85 | 140 | 1 | – | 70 | 55 | – | – | General purpose die |
| | | Si 10.0 | Chill cast | F | 95 | 185 | 2 | – | 80 | 60 | – | – | casting alloy |
| Al–Cu–Mg | (LM12) | Cu 10.0 | Chill cast | F | 155 | 185 | 1 | – | 85 | – | 0.9 | – | Castings to withstand |
| | | Mg 0.25 | Chill cast | T6 | 285 | 310 | – | – | 130 | 60 | – | – | high hydraulic pressure |
| Al–Cu | (L119) | Cu 5.0<br>Ni 1.5 | Sand cast | T6 | 200 | 225 | 2 | – | 90 | – | – | – | Sand castings for elevated temperature service |
| Al–Zn–Mg | (DTT)<br>5008B | Zn 5.3<br>Mg 0.6<br>Cr 0.5 | Sand cast | T4 | – | 220 | 5 | – | – | – | – | – | Colour anodizing alloy |
| Al–Cu–Si–Zn | (LM27) | Cu 2.0<br>Si 7.0 | Sand cast | F | 85 | 155 | 2 | – | 75 | – | – | – | Versatile general purpose alloy |
| | | | Chill cast | F | 100 | 180 | 3 | – | 80 | – | – | – | |
| Al–Si–Cu–Mg | (LM30) | Si 17.0 | Chill cast | F | 160 | 180 | 0.5 | – | 110 | – | – | – | Die castings with high |
| | | Cu 4.5 | Die cast | F | 240 | 275 | 1 | – | 120 | – | – | – | wear resistance, |
| | | Mg 0.6 | Die cast | O | 265 | 295 | 1 | – | – | – | – | – | especially automobile cylinder blocks |
| | (LM16) | Si 5.0 | Sand cast | T4 | 130 | 210 | 3 | 200 | 80 | 70 | 1.5 | – | Water-cooled cylinder |
| | | Cu 1.0<br>Mg 0.5 | Chill cast | T4 | 130 | 245 | 6 | 210 | 85 | 85 | 2.5 | – | heads and applications requiring leak-proof castings |
| | | | Sand cast | T6 | 245 | 255 | 1 | 215 | 100 | 60 | 1 | – | |
| | | | Chill cast | T6 | 275 | 310 | 2 | 225 | 110 | 70 | 1.5 | – | |

*continued overleaf*

**Table 3.2** *(continued)*

*Cast Alloys*

| Specification | Nominal composition % | Form | Condition | 0.2% Proof stress MPa | Tensile strength MPa | Elong,% on 50 mm (≥2.6 mm) or $5.65\sqrt{S_0}$ | Shear strength MPa | Brinell hardness ($P = 5D^2$) | Fatigue strength (unnotched) 500 MHz MPa | Impact energy J | Fracture toughness (MPa m$^{1/2}$) | Remarks |
|---|---|---|---|---|---|---|---|---|---|---|---|---|
| Al–Si–Mg–Mn (LM9) | Si 12.0 Mg 0.4 Mn 0.5 | Sand cast Chill cast | T5 T5 | 120 160 | 185 255 | 2 2.5 | 120 160 | 70 80 | 55* 70* | 1.5 2.5 | — — | Fluidity, corrosion resistance and high strength. Extensive use for low-pressure castings |
| (LM25) | Si 7.0 Mg 0.3 | Sand cast Chill cast | T6 T6 | 235 275 | 255 310 | 1 1 | 200 230 | 100 110 | 70* 85* | 0.7 1.5 | — — | The most widely used general purpose, high-strength casting alloy |
| | | Sand cast | F | 90 | 140 | 2.5 | — | 60 | — | — | — | |
| | | Chill cast | F | 90 | 180 | 4 | — | 60 | — | — | — | |
| | | Sand cast | T5 | 135 | 165 | 1.5 | — | 75 | 55 | — | — | |
| | | Chill cast | T5 | 165 | 220 | 2.5 | — | 85 | — | — | — | |
| | | Sand cast | T7 | 95 | 170 | 3 | — | 65 | — | — | — | |
| | | Chill cast | T7 | 100 | 230 | 8 | — | 70 | 75 | — | — | |
| | | Sand cast | T6 | 225 | 255 | 1 | — | 105 | 60 | — | — | |
| | | Chill cast | T6 | 240 | 310 | 3 | — | 105 | 95 | — | — | |
| Al–Cu–Mg–Ni (L35) (Y Alloy) | Cu 4.0 Mg 1.5 Ni 2.0 | Sand cast Chill cast | T6 T6 | 220 240 | 235 290 | 1 2 | — — | 115 115 | 80* 110* | 1.5 4.5 | — — | Highly stressed components operating at elevated temperatures |
| Al–Si–Cu–Mg–Zn (LM21) | Si 6.0 Cu 4.0 Mg 0.2 Zn 1.0 | Sand cast Chill cast | F F | 130 130 | 180 200 | 1 2 | — — | 85 90 | — — | — — | — — | General engineering applications, particular crankcases |
| Al–Si–Cu–Mg–Ni | Si 23.0 Cu 1.0 | Sand cast Chill cast | T5 T5 | 120 170 | 130 210 | 0.3 0.3 | — — | 120 120 | — — | — — | — — | Pistons for high performance internal combustion engines |
| (LM29) | Mg 1.0 Ni 1.0 | Sand cast Chill cast | T6 T6 | 120 170 | 130 210 | 0.3 0.3 | — — | 120 120 | — — | — — | — — | High performance piston alloy |

| | Composition | Casting | Temper | | | | | | | | Remarks |
|---|---|---|---|---|---|---|---|---|---|---|---|
| (LM28) | Si 19.0<br>Cu 1.5<br>Mg 1.0<br>Ni 1.0 | Chill cast | T5 | 170 | 190 | 0.5 | 120 | – | – | – | Low expansion piston alloy |
| | Si 11.0<br>Cu 1.0 | Sand cast | T6 | 120 | 130 | 0.5 | 120 | – | – | – | |
| | | Chill cast | T6 | 170 | 200 | 0.5 | 120 | – | – | – | |
| | | Chill cast | T5 | – | 220 | 1 | 105 | – | – | – | |
| (LM13) | Mg 1.0<br>Ni 1.0 | Sand cast | T6 | 190 | 200 | 0.5 | 115 | – | 85* | – | |
| | | Chill cast | T6 | 280 | 290 | 1 | 125 | 190 | 100* | 1.4 | |
| Lo-Ex | | Sand cast | T7 | 140 | 150 | 1 | 75 | – | – | – | |
| | | Chill cast | T7 | 200 | 210 | 1 | 75 | – | – | – | |
| (LM26) | Si 9.0<br>Cu 3.0<br>Mg 1.0<br>Ni 0.7 | Chill cast | T5 | 180 | 230 | 1 | 105 | – | – | 1.4 | Piston alloy |
| (3L52)<br>Al–Cu–Si–<br>Mg–Fe–Ni | Cu 2.0<br>Si 1.5<br>Mg 1.0<br>Fe 1.0<br>Ni 1.25 | Sand cast | T6 | 260 | 285 | 1 | 120 | – | 80 | – | Aircraft engine castings for elevated temperature service |
| | | Chill cast | T6 | 305 | 335 | 1 | 125 | – | – | – | |
| (3L51)<br>Al–Cu–Si–<br>Fe–Ni–Mg | Cu 1.5<br>Si 2.0<br>Fe 1.0<br>Ni 1.4<br>Mg 0.15 | Sand cast | T5 | 135 | 170 | 2.5 | 70 | – | – | – | Aircraft engine castings |
| | | Chill cast | T5 | 150 | 210 | 3.5 | 75 | – | – | – | |

*Fatigue Limit for 50 × 10$^6$ cycles.

M = as manufactured.

H111 = annealed.

H2, H4 } intermediate tempers.
H5, H6

H8 = fully hard temper.

(1) Special temper for maximum stress corrosion resistance (US designation T73).
(2) Special heat treatment for combination of properties (US designation T736).
(3) Special heat treatment for combination of properties (US designation T61).
(4) Special heat treatment for combination of properties (US designation T7351).

**Table 3.3**   ALUMINIUM AND ALUMINIUM ALLOYS – MECHANICAL PROPERTIES AT ELEVATED TEMPERATURES

| Material (specification) | Nominal composition % | Condition | Temp. °C | Time at temp. h | 0.2% Proof stress MPa | Tensile strength MPa | Elong. % on 50 mm or $5.65\sqrt{S_0}$ |
|---|---|---|---|---|---|---|---|
| | | | *Wrought Alloys* | | | | |
| Al (1095) | Al 99.95 | Rolled rod | H111 | 24 | – | – | 55 | 61 |
| | | | | 93 | – | – | 45 | 63 |
| | | | | 203 | – | – | 25 | 80 |
| | | | | 316 | – | – | 12 | 105 |
| | | | | 427 | – | – | 5 | 131 |
| (1200) | Al 99 | | H111 | 24 | 10 000 | 35 | 90 | 45 |
| | | | | 100 | 10 000 | 35 | 75 | 45 |
| | | | | 148 | 10 000 | 30 | 60 | 55 |
| | | | | 203 | 10 000 | 25 | 40 | 65 |
| | | | | 260 | 10 000 | 14 | 30 | 75 |
| | | | | 316 | 10 000 | 11 | 17 | 80 |
| | | | | 371 | 10 000 | 6 | 14 | 85 |
| | | | H14 | 24 | 10 000 | 115 | 125 | 20 |
| | | | | 100 | 10 000 | 105 | 110 | 20 |
| | | | | 148 | 10 000 | 85 | 90 | 22 |
| | | | | 203 | 10 000 | 50 | 65 | 25 |
| | | | | 260 | 10 000 | 17 | 30 | 75 |
| | | | | 316 | 10 000 | 11 | 17 | 80 |
| | | | | 371 | 10 000 | 6 | 14 | 85 |
| | | | H18 | 24 | 10 000 | 150 | 165 | 15 |
| | | | | 100 | 10 000 | 125 | 150 | 15 |
| | | | | 148 | 10 000 | 95 | 125 | 20 |
| | | | | 203 | 10 000 | 30 | 40 | 65 |
| | | | | 260 | 10 000 | 14 | 30 | 75 |
| | | | | 316 | 10 000 | 11 | 17 | 80 |
| | | | | 371 | 10 000 | 6 | 14 | 85 |
| Al–Mn (3103) | Mn 1.25 | | H111 | 24 | 10 000 | 40 | 110 | 40 |
| | | | | 100 | 10 000 | 37 | 90 | 43 |
| | | | | 148 | 10 000 | 34 | 75 | 47 |
| | | | | 203 | 10 000 | 30 | 60 | 60 |
| | | | | 260 | 10 000 | 25 | 40 | 65 |
| | | | | 316 | 10 000 | 17 | 30 | 70 |
| | | | | 371 | 10 000 | 14 | 20 | 70 |
| | | | H14 | 24 | 10 000 | 145 | 150 | 16 |
| | | | | 100 | 10 000 | 130 | 145 | 16 |
| | | | | 148 | 10 000 | 110 | 125 | 16 |
| | | | | 203 | 10 000 | 60 | 95 | 20 |
| | | | | 260 | 10 000 | 30 | 50 | 60 |
| | | | | 316 | 10 000 | 17 | 30 | 70 |
| | | | | 371 | 10 000 | 14 | 20 | 70 |
| | | | H18 | 24 | 10 000 | 185 | 200 | 10 |
| | | | | 148 | 10 000 | 110 | 155 | 11 |
| | | | | 203 | 10 000 | 60 | 95 | 18 |
| | | | | 260 | 10 000 | 30 | 50 | 60 |
| | | | | 316 | 10 000 | 17 | 30 | 70 |
| | | | | 371 | 10 000 | 14 | 20 | 70 |
| Al–Mg (5050) | Mg 1.4 | | H111 | 24 | 10 000 | 55 | 145 | – |
| | | | | 100 | 10 000 | 55 | 145 | – |
| | | | | 148 | 10 000 | 55 | 130 | – |
| | | | | 203 | 10 000 | 50 | 95 | – |
| | | | | 260 | 10 000 | 40 | 60 | – |
| | | | | 316 | 10 000 | 30 | 40 | – |
| | | | | 371 | 10 000 | 20 | 30 | – |

**Table 3.3**   *(continued)*

| Material (specification) | Nominal composition % | Condition | Temp. °C | Time at temp. h | 0.2% Proof stress MPa | Tensile strength MPa | Elong. % on 50 mm or $5.65\sqrt{S_0}$ |
|---|---|---|---|---|---|---|---|
| Al–Mg *(cont.)* | | H14 | 24 | 10 000 | 165 | 195 | – |
| | | | 100 | 10 000 | 165 | 195 | – |
| | | | 148 | 10 000 | 150 | 165 | – |
| | | | 203 | 10 000 | 50 | 95 | – |
| | | | 260 | 10 000 | 40 | 60 | – |
| | | | 316 | 10 000 | 35 | 40 | – |
| | | | 371 | 10 000 | 20 | 30 | – |
| | | H18 | 24 | 10 000 | 200 | 220 | – |
| | | | 100 | 10 000 | 200 | 215 | – |
| | | | 148 | 10 000 | 175 | 180 | – |
| | | | 203 | 10 000 | 60 | 95 | – |
| | | | 260 | 10 000 | 40 | 60 | – |
| | | | 316 | 10 000 | 35 | 40 | – |
| | | | 371 | 10 000 | 20 | 30 | – |
| Al–Mg–Cr (5052) | Mg 2.25 Cr 0.25 | H111 | 24 | 10 000 | 90 | 195 | 30 |
| | | | 100 | 10 000 | 90 | 190 | 35 |
| | | | 148 | 10 000 | 90 | 165 | 50 |
| | | | 203 | 10 000 | 75 | 125 | 65 |
| | | | 260 | 10 000 | 50 | 80 | 80 |
| | | | 316 | 10 000 | 35 | 50 | 100 |
| | | | 371 | 10 000 | 20 | 35 | 130 |
| | | H14 | 24 | 10 000 | 215 | 260 | 14 |
| | | | 100 | 10 000 | 205 | 260 | 16 |
| | | | 148 | 10 000 | 185 | 215 | 25 |
| | | | 203 | 10 000 | 105 | 155 | 40 |
| | | | 260 | 10 000 | 50 | 80 | 80 |
| | | | 316 | 10 000 | 35 | 50 | 100 |
| | | | 317 | 10 000 | 20 | 35 | 130 |
| | | H18 | 24 | 10 000 | 255 | 290 | 8 |
| | | | 100 | 10 000 | 255 | 285 | 9 |
| | | | 148 | 10 000 | 200 | 235 | 20 |
| | | | 203 | 10 000 | 105 | 155 | 40 |
| | | | 260 | 10 000 | 50 | 80 | 80 |
| | | | 316 | 10 000 | 35 | 50 | 100 |
| | | | 371 | 10 000 | 20 | 35 | 130 |
| (5154) | Mg 3.5 Cr 0.25 | H111 | 24 | 10 000 | 125 | 240 | 25 |
| | | | 100 | 10 000 | 125 | 240 | 30 |
| | | | 148 | 10 000 | 125 | 195 | 40 |
| | | | 203 | 10 000 | 95 | 145 | 55 |
| | | | 260 | 10 000 | 60 | 110 | 70 |
| | | | 316 | 10 000 | 40 | 70 | 100 |
| | | | 371 | 10 000 | 30 | 40 | 130 |
| | | H14 | 24 | 10 000 | 225 | 290 | 12 |
| | | | 100 | 10 000 | 220 | 285 | 16 |
| | | | 148 | 10 000 | 195 | 235 | 25 |
| | | | 203 | 10 000 | 110 | 175 | 35 |
| | | | 260 | 10 000 | 60 | 110 | 70 |
| | | | 316 | 10 000 | 40 | 70 | 100 |
| | | | 371 | 10 000 | 30 | 40 | 130 |
| | | H18 | 24 | 10 000 | 270 | 330 | 8 |
| | | | 100 | 10 000 | 255 | 310 | 13 |
| | | | 148 | 10 000 | 220 | 270 | 20 |
| | | | 203 | 10 000 | 105 | 155 | 35 |
| | | | 260 | 10 000 | 60 | 110 | 70 |
| | | | 316 | 10 000 | 40 | 70 | 100 |
| | | | 371 | 10 000 | 30 | 40 | 130 |

*continued overleaf*

**Table 3.3**    (*continued*)

| Material (specification) | Nominal composition % | Condition | | Temp. °C | Time at temp. h | 0.2% Proof stress MPa | Tensile strength MPa | Elong. % on 50 mm or 5.65$\sqrt{S_0}$ |
|---|---|---|---|---|---|---|---|---|
| Al–Mg–Mn (5056A) | Mg 5.0 Mn 0.3 | As extruded | F | 20 | 1 000 | 145 | 300 | 25 |
| | | | | 50 | 1 000 | 145 | 300 | 27 |
| | | | | 100 | 1 000 | 145 | 300 | 32 |
| | | | | 150 | 1 000 | 135 | 245 | 45 |
| | | | | 200 | 1 000 | 111 | 215 | 56 |
| | | | | 250 | 1 000 | 75 | 130 | 77 |
| | | | | 300 | 1 000 | 50 | 95 | 100 |
| | | | | 350 | 1 000 | 20 | 60 | 140 |
| Al–Mg–Si (6063) | Mg 0.7 Si 0.4 | | T6 | 24 | 10 000 | 215 | 240 | 18 |
| | | | | 100 | 10 000 | 195 | 215 | 15 |
| | | | | 148 | 10 000 | 135 | 145 | 20 |
| | | | | 203 | 10 000 | 45 | 60 | 40 |
| | | | | 260 | 10 000 | 25 | 30 | 75 |
| | | | | 316 | 10 000 | 17 | 20 | 80 |
| | | | | 371 | 10 000 | 14 | 17 | 105 |
| (6082) | Mg 0.6 Si 1.0 Cr 0.25 | | T6 | 24 | 10 000 | 230 | 330 | 17 |
| | | | | 100 | 10 000 | 270 | 290 | 19 |
| | | | | 148 | 10 000 | 175 | 185 | 22 |
| | | | | 203 | 10 000 | 65 | 80 | 40 |
| | | | | 206 | 10 000 | 35 | 45 | 50 |
| | | | | 316 | 10 000 | 30 | 35 | 50 |
| | | | | 371 | 10 000 | 25 | 30 | 50 |
| (6061) | Mg 1.0 Si 0.6 Cu 0.25 Cr 0.25 | | T6 | 24 | 10 000 | 275 | 310 | 17 |
| | | | | 100 | 10 000 | 260 | 290 | 18 |
| | | | | 148 | 10 000 | 213 | 235 | 20 |
| | | | | 203 | 10 000 | 105 | 130 | 28 |
| | | | | 260 | 10 000 | 35 | 50 | 60 |
| | | | | 316 | 10 000 | 17 | 30 | 85 |
| | | | | 371 | 10 000 | 14 | 20 | 95 |
| Al–Cu–Mn (2219) | Cu 6.0 Mn 0.25 | Forgings | T6 | 20 | 100 | 230 | 385 | 8 |
| | | | | 100 | 100 | – | 365 | – |
| | | | | 150 | 100 | 220 | 325 | – |
| | | | | 200 | 100 | 185 | 280 | – |
| | | | | 250 | 100 | 135 | 205 | – |
| | | | | 300 | 100 | 110 | 145 | – |
| | | | | 350 | 100 | 45 | 70 | – |
| | | | | 400 | 100 | 20 | 30 | – |
| Al–Cu–Pb–Bi (2011) | Cu 5.5 Pb 0.5 Bi 0.5 | | T4 | 24 | 10 000 | 295 | 375 | 15 |
| | | | | 100 | 10 000 | 235 | 320 | 16 |
| | | | | 148 | 10 000 | 130 | 195 | 25 |
| | | | | 203 | 10 000 | 75 | 110 | 35 |
| | | | | 260 | 10 000 | 30 | 45 | 45 |
| | | | | 316 | 10 000 | 14 | 25 | 90 |
| | | | | 371 | 10 000 | 11 | 17 | 125 |
| Al–Cu–Mg–Mn (2017) | Cu 4.0 Mg 0.5 Mn 0.5 | | T4 | 24 | 10 000 | 275 | 430 | 22 |
| | | | | 100 | 10 000 | 255 | 385 | 18 |
| | | | | 148 | 10 000 | 205 | 274 | 16 |
| | | | | 203 | 10 000 | 115 | 150 | 28 |
| | | | | 260 | 10 000 | 65 | 80 | 45 |
| | | | | 316 | 10 000 | 35 | 45 | 95 |
| | | | | 371 | 10 000 | 25 | 30 | 100 |
| (2024) | Cu 4.5 Mg 1.5 Mn 0.6 | | T4 | 24 | 10 000 | 340 | 470 | 19 |
| | | | | 100 | 10 000 | 305 | 422 | 17 |
| | | | | 148 | 10 000 | 245 | 295 | 17 |
| | | | | 203 | 10 000 | 145 | 180 | 22 |
| | | | | 260 | 10 000 | 65 | 95 | 45 |
| | | | | 316 | 10 000 | 35 | 50 | 75 |
| | | | | 371 | 10 000 | 25 | 35 | 100 |

**Table 3.3**  *(continued)*

| Material (specification) | Nominal composition % | Condition | | Temp. °C | Time at temp. h | 0.2% Proof stress MPa | Tensile strength MPa | Elong. % on 50 mm or $5.65\sqrt{S_0}$ |
|---|---|---|---|---|---|---|---|---|
| *Wrought alloys* | | | | | | | | |
| Al–Cu–Mg–Si–Mn (2014) | Cu 4.4 Mg 0.4 Si 0.8 Mn 0.8 | | T6 | 24 | 10 000 | 415 | 485 | 13 |
| | | | | 100 | 10 000 | 385 | 455 | 14 |
| | | | | 148 | 10 000 | 275 | 325 | 15 |
| | | | | 203 | 10 000 | 80 | 125 | 35 |
| | | | | 260 | 10 000 | 60 | 75 | 45 |
| | | | | 316 | 10 000 | 35 | 45 | 64 |
| | | | | 371 | 10 000 | 25 | 30 | 20 |
| | | Forgings | T6 | 20 | 100 | 415* | 480 | 10 |
| | | | | 100 | 100 | 410 | 465 | – |
| | | | | 150 | 100 | 400 | 430 | – |
| | | | | 200 | 100 | 260 | 295 | – |
| | | | | 250 | 100 | 85 | 110 | – |
| | | | | 300 | 100 | 45 | 70 | – |
| | | | | 350 | 100 | 35 | 50 | – |
| Al–Cu–Mg–Ni (2618) | Cu 2.2 Mg 1.5 Ni 1.2 Fe 1.0 | Forgings | T6 | 20 | 100 | 325* | 430 | 8 |
| | | | | 150 | 100 | 340 | 440 | – |
| | | | | 200 | 100 | 260 | 300 | – |
| | | | | 250 | 100 | 170 | 210 | – |
| | | | | 300 | 100 | 70 | 115 | – |
| | | | | 350 | 100 | 30 | 50 | – |
| | | | | 400 | 100 | 20 | 30 | – |
| (2031) | Cu 2.2 Mg 1.5 Ni 1.2 Fe 1.0 Si 0.8 | Forgings | T6 | 20 | 100 | 325* | 430 | 13 |
| | | | | 100 | 100 | 310 | 400 | – |
| | | | | 200 | 100 | 255 | 310 | – |
| | | | | 250 | 100 | 110 | 155 | – |
| | | | | 300 | 100 | 45 | 75 | – |
| | | | | 350 | 100 | 30 | 40 | – |
| Al–Si–Cu–Mg–Ni (4032) | Si 12.2 Cu 0.9 Mg 1.1 Ni 0.9 | Forgings | T6 | 24 | 10 000 | 320 | 380 | 9 |
| | | | | 100 | 10 000 | 305 | 345 | 9 |
| | | | | 148 | 10 000 | 225 | 255 | 9 |
| | | | | 203 | 10 000 | 60 | 90 | 30 |
| (4032) | | | | 260 | 10 000 | 35 | 55 | 50 |
| | | | | 316 | 10 000 | 20 | 35 | 70 |
| | | | | 371 | 10 000 | 14 | 25 | 90 |
| Al–Zn–Mg–Cu (7075) | Zn 5.6 Cu 1.6 Mg 2.5 Cr 0.3 | | T6 | 24 | 10 000 | 505 | 570 | 11 |
| | | | | 100 | 10 000 | 430 | 455 | 15 |
| | | | | 148 | 10 000 | 145 | 175 | 30 |
| | | | | 203 | 10 000 | 80 | 95 | 60 |
| | | | | 260 | 10 000 | 60 | 75 | 65 |
| | | | | 316 | 10 000 | 45 | 60 | 80 |
| | | | | 371 | 10 000 | 30 | 45 | 65 |
| *Cast alloys* | | | | | | | | |
| Al–Mg (LM 5) | Mg 5.0 Mn 0.5 | Sand cast | F | 20 | 1 000 | 95 | 160 | 4 |
| | | | | 100 | 1 000 | 100 | 160 | 3 |
| | | | | 200 | 1 000 | 95 | 130 | 3 |
| | | | | 300 | 1 000 | 55 | 95 | 4 |
| | | | | 400 | 1 000 | 15 | 30 | 4 |
| (LM 10) | Mg 10.0 | Sand cast | T4 | 20 | 1 000 | 180 | 340 | 16 |
| | | | | 100 | 1 000 | 205 | 350 | 10 |
| | | | | 150 | 1 000 | 154 | 270 | 0 |
| | | | | 200 | 1 000 | 105 | 185 | 42 |
| | | | | 300 | 1 000 | 40 | 90 | 85 |
| | | | | 400 | 1 000 | 11 | 45 | 100 |
| Al–Si (LM 18) | Si 5.0 | Pressure die cast | F | 24 | 10 000 | 110 | 205 | 9 |
| | | | | 100 | 10 000 | 110 | 175 | 9 |
| | | | | 148 | 10 000 | 103 | 135 | 10 |
| | | | | 203 | 10 000 | 80 | 110 | 17 |
| | | | | 260 | 10 000 | 40 | 55 | 23 |

*continued overleaf*

**Table 3.3** (continued)

| Material (specification) | Nominal composition % | Condition | | Temp. °C | Time at temp. h | 0.2% Proof stress MPa | Tensile strength MPa | Elong. % on 50 mm or 5.65√$S_0$ |
|---|---|---|---|---|---|---|---|---|
| (LM 6) | Si 12.0 | Pressure die cast | F | 24 | 10 000 | 145 | 270 | 2 |
| | | | | 100 | 10 000 | 145 | 225 | $2\frac{1}{2}$ |
| | | | | 148 | 10 000 | 125 | 185 | 3 |
| | | | | 206 | 10 000 | 105 | 150 | 7 |
| | | | | 260 | 10 000 | 40 | 75 | 13 |
| Al–Si–Cu (LM 4) | Si 5.0 Cu 3.0 Mn 0.5 | Sand cast | F | 20 | 1000 | 95* | 155 | 2 |
| | | | | 100 | 1000 | 140 | 180 | 2 |
| | | | | 200 | 1000 | 110 | 135 | 2 |
| | | | | 300 | 1000 | 40 | 60 | 12 |
| | | | | 400 | 1000 | 20 | 30 | 27 |
| Al–Si–Mg (LM 25) | Si 5.0 Mg 0.5 | Chill cast | T6 | 20 | 1000 | 270* | 325 | 2 |
| | | | | 100 | 1000 | 255 | 290 | 2 |
| | | | | 200 | 1000 | 60 | 90 | 25 |
| | | | | 300 | 1000 | 25 | 40 | 65 |
| | | | | 400 | 1000 | 12 | 25 | 65 |
| Al–Cu–Mg–Ni (4L 35) | Cu 4.0 Mg 1.5 Ni 2.0 | Sand cast | T6 | 20 | 1000 | 200* | 275 | $\frac{1}{2}$ |
| | | | | 100 | 1000 | 255 | 325 | $\frac{1}{2}$ |
| | | | | 200 | 1000 | 150 | 135 | $\frac{1}{2}$ |
| | | | | 300 | 1000 | 30 | 55 | 32 |
| | | | | 400 | 1000 | 15 | 40 | 60 |
| Al–Si–Ni–Cu–Mg (LM 13) | Si 12.0 Ni 2.5 Cu 1.0 Mg 1.0 | Chill cast | T6 | 20 | 1000 | 275* | 285 | $\frac{1}{2}$ |
| | | | | 100 | 1000 | 280 | 320 | $\frac{1}{2}$ |
| | | | | 200 | 1000 | 110 | 165 | $\frac{1}{2}$ |
| | | | | 300 | 1000 | 30 | 60 | 15 |
| | | | | 400 | 1000 | 15 | 35 | 25 |
| | | Chill cast Special | T6 | 20 | 1000 | 200* | 275 | 1 |
| | | | | 100 | 1000 | 195 | 250 | 1 |
| | | | | 200 | 1000 | 110 | 170 | 3 |
| | | | | 300 | 1000 | 35 | 60 | 15 |
| | | | | 400 | 1000 | 15 | 35 | 50 |

*0.1% Proof stress.

**Table 3.4** ALUMINIUM AND ALUMINIUM ALLOYS – MECHANICAL PROPERTIES AT LOW TEMPERATURES

| Material (specification) | Nominal composition % | Condition | | Temp. °C | 0.2% Proof stress MPa | Tensile strength MPa | Elong.% on 50 mm or 50 mm | Reduction in area % | Fracture toughness MPa m$^{1/2}$ |
|---|---|---|---|---|---|---|---|---|---|
| Al (1200) | Al 9.0 | Rolled and drawn rod | H111 | 24 | 34 | 90 | 42.5 | 76.4 | – |
| | | | | −28 | 34 | 95 | 43.0 | 76.4 | – |
| | | | | −80 | 37 | 100 | 47.5 | 77.0 | – |
| | | | | −196 | 43 | 170 | 56 | 74.4 | – |
| | | | H18 | 24 | 140 | 155 | 16 | 59.8 | – |
| | | | | −28 | 144 | 155 | 152 | 59.4 | – |
| | | | | −80 | 147 | 165 | 18.0 | 65.3 | – |
| | | | | −196 | 165 | 225 | 35.2 | 67.0 | – |
| Al–Mn (3103) | Mn 1.25 | Rolled and drawn rod | H111 | 24 | 40 | 110 | 43.0 | 80.6 | – |
| | | | | −28 | 40 | 115 | 44.0 | 80.6 | – |
| | | | | −80 | 50 | 130 | 45.0 | 79.9 | – |
| | | | | −196 | 60 | 220 | 48.8 | 71.2 | – |

**Table 3.4**  (*continued*)

| Material (specification) | Nominal composition % | Condition | | Temp. °C | 0.2% Proof stress MPa | Tensile strength MPa | Elong.% on 50 mm or 50 mm | Reduction in area % | Fracture toughness MPa m$^{1/2}$ |
|---|---|---|---|---|---|---|---|---|---|
| | | Rolled and drawn rod | H18 | 24 | 180 | 195 | 15.0 | 63.5 | – |
| | | | | 28 | 185 | 205 | 15.0 | 64.4 | – |
| | | | | −80 | 195 | 215 | 16.5 | 66.5 | – |
| | | | | −196 | 220 | 290 | 32.0 | 62.3 | – |
| Al–Mg (5052) | Mg 2.5 Cr 0.25 | Rolled and drawn rod | H111 | 24 | 97 | 199 | 33.2 | 72.0 | – |
| | | | | −28 | 99 | 201 | 35.8 | 74.2 | – |
| | | | | −80 | 97 | 210 | 40.8 | 76.4 | – |
| | | | | −196 | 115 | 330 | 50.0 | 69.0 | – |
| | | | H18 | 24 | 235 | 275 | 16.6 | 59.1 | – |
| | | | | −28 | 230 | 280 | 18.3 | 63.2 | – |
| | | | | −80 | 236 | 290 | 20.6 | 64.5 | – |
| | | | | −196 | 275 | 400 | 30.9 | 57.4 | – |
| (5154) | Mg 3.5 Cr 0.25 | Sheet | H111 | 26 | 115 | 240 | 28 | 66 | – |
| | | | | −28 | 115 | 240 | 32 | 72 | – |
| | | | | −80 | 115 | 250 | 35 | 73 | – |
| | | | | −196 | 135 | 350 | 42 | 60 | – |
| | | | H18 | 26 | 275 | 330 | 9 | – | – |
| | | | | −80 | 280 | 340 | 14 | – | – |
| | | | | −196 | 325 | 455 | 30 | – | – |
| | | | | −253 | 370 | 645 | 35 | – | – |
| (5056A) | Mg 5.0 Mn 0.2 | Plate | H111 | 20 | 130 | 290 | 30.5 | 32.0 | – |
| | | | | −75 | 130 | 290 | 38.2 | 48.2 | – |
| | | | | −196 | 145 | 420 | 50.0 | 36.2 | – |
| Al–Mg–Si (6063) | Mg 0.7 Si 0.4 | Extrusion | T4 | 26 | 90 | 175 | 32 | 78 | – |
| | | | | −28 | 105 | 190 | 33 | 75 | – |
| | | | | −80 | 115 | 200 | 36 | 75 | – |
| | | | | −196 | 115 | 260 | 42 | 73 | – |
| | | Extrusion | T6 | 26 | 215 | 240 | 16 | 36 | – |
| | | | | −28 | 220 | 250 | 16 | 36 | – |
| | | | | −80 | 225 | 260 | 17 | 38 | – |
| | | | | −196 | 250 | 330 | 21 | 40 | – |
| Al–Mg–Si-Cr (6151) | Mg 0.7 Si 1.0 Cr 0.25 | Forging | T6 | 24 | 300 | 320 | 15.2 | 38.8 | – |
| | | | | −28 | 310 | 352 | 12.0 | 34.0 | – |
| | | | | −80 | 305 | 330 | 14.9 | 38.7 | – |
| | | | | −196 | 330 | 385 | 18.3 | 34.7 | – |
| Al–Mg–Si–Cu–Cr (6061) | Mg 1.0 Si 0.6 Cu 0.25 Cr 0.25 | Rolled and drawn rod | T6 | 24 | 270 | 315 | 21.8 | 56.4 | – |
| | | | | −28 | 280 | 330 | 21.5 | 52.5 | – |
| | | | | −80 | 290 | 345 | 22.5 | 53.7 | – |
| | | | | −196 | 315 | 425 | 26.5 | 46.5 | – |
| Al–Cu–Mg–Mn (2024) | Cu 4.5 Mg 1.5 Mn 0.6 | Rolled and drawn rod | T4 | 24 | 300 | 480 | 23.3 | 31.8 | – |
| | | | | −28 | 305 | 500 | 24.4 | 33.1 | – |
| | | | | −80 | 320 | 510 | 25.3 | 30.8 | – |
| | | | | −196 | 400 | 615 | 26.7 | 26.3 | – |
| | | Rolled and drawn rod | T8 | 24 | 400 | 500 | 14.5 | 25.8 | – |
| | | | | −28 | 405 | 502 | 12.7 | 21.5 | – |
| | | | | −80 | 415 | 514 | 13.3 | 22.0 | – |
| | | | | −196 | 460 | 605 | 14.0 | 19.7 | – |
| Al–Cu–Si–Mg–Mn (2014) | Cu 4.5 Si 0.8 Mg 0.5 Mn 0.8 | Rod | T4 | 26 | 290 | 430 | 20 | 28 | – |
| | | | | −28 | 290 | 440 | 22 | 28 | – |
| | | | | −80 | 302 | 440 | 22 | 26 | – |
| | | | | −196 | 380 | 545 | 20 | 20 | – |
| | | Rod | T6 | 26 | 415 | 485 | 13 | 31 | – |
| | | | | −28 | 415 | 485 | 13 | 29 | – |
| | | | | −80 | 420 | 495 | 14 | 28 | – |
| | | | | −196 | 470 | 565 | 14 | 26 | – |

*continued overleaf*

**Table 3.4**   (*continued*)

| Material (specification) | Nominal composition % | Condition | | Temp. °C | 0.2% Proof stress MPa | Tensile strength MPa | Elong.% on 50 mm or 50 mm | Reduction in area % | Fracture toughness MPa m$^{1/2}$ |
|---|---|---|---|---|---|---|---|---|---|
| | | Forging T6 | | 26 | 410 | 465 | 12 | 24 | – |
| | | | | −80 | 460 | 510 | 14 | 24 | – |
| | | | | −196 | 530 | 610 | 11 | 22 | – |
| | | | | −253 | 590 | 715 | 7 | 22 | – |
| (2090) | Cu 2.7 Li 2.3 Zr 0.12 | Plate (12.5 mm) | T81 | 27 | 535 | 565 | 11 | – | 34 |
| | | | | −196 | 600 | 715 | 13.5 | – | 57 |
| | | | | −269 | 615 | 820 | 17.5 | – | 72 |
| (2091) | Cu 2.1 Li 2.0 Mg 1.50 Zr 0.1 | Plate (38 mm) | T8 | 27 | 440 | 480 | 6 | – | 24 |
| | | | | −73 | 460 | 495 | 7 | – | 32 |
| | | | | −196 | 495 | 565 | 10 | – | 32 |
| | | | | −269 | 550 | 630 | 7 | – | 32 |
| Al–Zn– Mg–Cu (7075) | Zn 5.6 Mg 2.5 Cu 1.6 | Rolled and drawn rod | T6 | 24 | 485 | 560 | 15.0 | 29.1 | – |
| | | | | −28 | 490 | 570 | 15.3 | 26.2 | – |
| | | | | −80 | 505 | 590 | 15.3 | 23.6 | – |
| | | | | −196 | 570 | 670 | 16.0 | 20.1 | – |

H111 = Annealed. H18 = Fully hard temper. T4 = Solution treated and naturally aged. T6 = Solution treated and precipitation treated.

**Table 3.5**   ALUMINIUM ALLOYS – CREEP DATA

| Material (specification) | Nominal composition % | Condition | | Temp. °C | Stress MPa | Minimum creep rate % per 1000 h | Total extension % in 1000 h |
|---|---|---|---|---|---|---|---|
| Al (1080) | 99.8 | Sheet | H111 | 20 | 24.1 | 0.005 | 0.39 |
| | | | | 20 | 27.6 | 0.045 | 1.28 |
| | | | | 80 | 7.0 | 0.005 | 0.045 |
| | | | | 80 | 8.3 | 0.01 | 0.065 |
| | | | | 250 | 1.4 | 0.005 | 0.047 |
| | | | | 250 | 2.1 | 0.01 | 0.047 |
| | | | | 250 | 2.8 | 0.015 | 0.052 |
| | | | | 250 | 4.1 | 0.055 | 0.152 |
| Al–Mg (5052) | Mg | Sheet | H111 | 80 | 45 | 0.005 | 0.085 |
| (LM 5) | Mg 5.6 | Cast | | 100 | 110 | 0.055 | 0.33 |
| | | | | 100 | 115 | 0.17 | 0.57 |
| | | | | 100 | 125 | 0.21 | 1.19 |
| | | | | 200 | 30 | 0.08 | 0.21 |
| | | | | 200 | 45 | 0.20 | 0.39 |
| | | | | 200 | 60 | 0.62 | 0.92 |
| | | | | 300 | 3.90 | 0.045 | 0.10 |
| | | | | 300 | 7.7 | 0.12 | 0.25 |
| | | | | 300 | 15 | 0.35 | 0.60 |
| (LM 10) | Mg 10 | Cast | | 100 | 40 | 0.013 | 0.126 |
| | | | | 100 | 55 | 0.022 | 0.107 |
| | | | | 100 | 75 | 0.046 | 0.174 |
| | | | | 150 | 7.5 | 0.126 | 0.413 |
| | | | | 150 | 15 | 0.147 | 0.647 |
| | | | | 200 | 7.5 | 0.107 | 0.341 |
| | | | | 200 | 15 | 0.273 | 0.658 |
| Al–Cu | Cu 4 | Cast | | 205 | 17 | 0.04 | – |
| | | | | 205 | 34 | 0.09 | – |
| | | | | 205 | 51 | 0.14 | – |

**Table 3.5**   *(continued)*

| Material (specification) | Nominal composition % | Condition | Temp. °C | Stress MPa | Minimum creep rate % per 1000 h | Total extension % in 1000 h |
|---|---|---|---|---|---|---|
| | | | 205 | 70 | 0.69 | – |
| | | | 315 | 8.90 | 0.13 | – |
| | | | 315 | 13.1 | 0.29 | – |
| | Cu 10 | Cast | 205 | 34 | 0.01 | – |
| | | | 205 | 68 | 0.11 | – |
| | | | 315 | 8.90 | 0.12 | – |
| | | | 315 | 13.1 | 0.43 | – |
| | | | 315 | 17 | 0.99 | – |
| Al–Si (LM 13) | Si 13 Ni 1.7 Mg 1.3 | Sandcast (modified) | 100 | 45 | 0.016 | 0.190 |
| | | | 100 | 60 | 0.06 | 0.675 |
| | | | 200 | 15 | 0.016 | 0.096 |
| | | | 200 | 23 | 0.054 | 0.179 |
| | | | 200 | 30 | 0.14 | 0.432 |
| | | | 300 | 3.8 | 0.013 | 0.026 |
| | | | 300 | 7.7 | 0.047 | 0.098 |
| | | | 300 | 15 | 0.223 | 0.428 |
| Al–Mn (3103) | Mn 1.25 | Extruded rod | 200 | 15 | 0.001 | – |
| | | | 200 | 31 | 0.022 | – |
| | | | 200 | 34.8 | 0.040 | – |
| | | | 200 | 38.6 | 0.060 | – |
| | | | 200 | 42.5 | 0.13 | – |
| | | | 200 | 46 | 0.15 | – |
| | | | 200 | 54 | 0.73 | – |
| | | | 300 | 7.5 | 0.007 | – |
| | | | 300 | 15 | 0.39 | – |
| Al–Cu–Si (2025) | Cu 4 Si 0.8 | Extruded T4 | 150 | 90 | 0.03 | 0.340 |
| | | | 150 | 125 | 0.045 | 0.395 |
| | | | 150 | 155 | 0.325 | 0.722 |
| | | | 200 | 30 | 0.035 | 0.107 |
| | | | 200 | 45 | 0.1 | 0.204 |
| | | | 200 | 60 | 0.040 | 0.700 |
| | | | 250 | 15 | 0.02 | 0.156 |
| | | | 250 | 23 | 0.07 | 0.176 |
| | | | 250 | 30 | 2.36 | – |
| Al–Cu–Mg–Mn (2024) | Cu 4.5 Mg 1.5 Mn 0.6 | Clad sheet T4 | 35 | 415 | 10.0 | – |
| | | | 100 | 344 | 1.0 | – |
| | | | 100 | 385 | 10.0 | – |
| | | | 150 | 276 | 1.0 | – |
| | | | 150 | 327 | 10.0 | – |
| | | | 190 | 140 | 1.0 | – |
| | | | 190 | 200 | 10.0 | – |
| | | Clad sheet T6 | 35 | 424 | 1.0 | – |
| | | | 35 | 430 | 10.0 | – |
| | | | 100 | 347 | 1.0 | – |
| | | | 100 | 363 | 10.0 | – |
| | | | 150 | 242 | 1.0 | – |
| | | | 150 | 289 | 10.0 | – |
| | | | 190 | 117 | 1.0 | – |
| | | | 190 | 193 | 10.0 | – |
| Al–Cu–Mg–Ni (2218) | Cu 4 Mg 1.5 Ni 2.2 | Forged T4 | 100 | 193 | 0.01 | 0.394 |
| | | | 100 | 232 | 0.02 | 0.440 |
| | | | 100 | 270 | 0.04 | 0.835 |
| | | | 200 | 77 | 0.028 | 0.173 |
| | | | 200 | 108 | 0.16 | 0.345 |
| | | | 300 | 7 | 0.037 | 0.078 |
| | | | 300 | 15 | 0.5 | 0.640 |
| | | | 400 | 1.5 | 0.05 | 0.110 |

*continued overleaf*

**Table 3.5**    *(continued)*

| Material (specification) | Nominal composition % | Condition | Temp. °C | Stress MPa | Minimum creep rate % per 1000 h | Total extension % in 1000 h |
|---|---|---|---|---|---|---|
| | | Cast T4 | 200 | 77 | 0.01 | 0.153 |
| | | | 200 | 116 | 0.08 | 0.287 |
| | | | 300 | 7 | 0.018 | 0.072 |
| | | | 300 | 15 | 0.08 | 0.151 |
| | | | 400 | 1.50 | 0.06 | 0.132 |
| Al–Cu–Mg–Zn (7075) | Zn 5.6 Cu 1.6 Mg 2.5 | Clad sheet T6 | 35 | 430 | 0.1 | – |
| | | | 35 | 480 | 1.0 | – |
| | | | 35 | 495 | 10.0 | – |
| | | | 100 | 295 | 0.1 | – |
| | | | 100 | 355 | 1.0 | – |
| | | | 100 | 370 | 10.0 | – |
| | | | 150 | 70 | 0.1 | – |
| | | | 150 | 170 | 1.0 | – |
| | | | 150 | 245 | 10.0 | – |
| | | | 190 | 45 | 0.1 | – |
| | | | 190 | 75 | 1.0 | – |
| | | | 190 | 125 | 10.0 | – |
| Al–Mg–Si–Mn (6351) | Mg 0.7 Si 1.0 Mn 0.6 | Extruded rod | 100 | 193 | 0.007 | – |
| | | | 100 | 201 | 0.010 | – |
| | | | 100 | 232 | 0.11 | – |
| | | | 100 | 255 | 1.6 | – |
| | | | 150 | 93 | 0.0087 | – |
| | | | 150 | 108 | 0.023 | – |
| | | | 150 | 154 | 0.22 | – |
| | | | 200 | 31 | 0.011 | – |
| | | | 200 | 46 | 0.040 | – |
| | | | 200 | 62 | 0.13 | – |
| | | | 200 | 77 | 0.28 | – |

H111 = Annealed.
T4 = Solution treated and naturally aged, will respond to precipitation treatment.
T6 = Solution treated and artificially aged.

**Table 3.6**    ALUMINIUM ALLOYS – FATIGUE STRENGTH AT VARIOUS TEMPERATURES

| Material (specification) | Nominal composition % | Condition | Temp. °C | Endurance (unnotched) MPa | MHz | Remarks |
|---|---|---|---|---|---|---|
| Al–Mg (5056) | Mg 5.0 | Extruded | −65 | 184 | 20 | Rotating beam |
| | | | −35 | 164 | | |
| | | | +20 | 133 | | |
| | Mg 7.0 | Extruded rod | −65 | 182 | 20 | Rotating beam |
| | | | −35 | 178 | | |
| | | | +20 | 173 | | |
| (LM 10) | Mg 10.0 | Sand cast (oil quenched) | 20 | 93 | 30 | Rotating beam |
| | | | 150 | 77 | | |
| | | | 200 | 40 | | |
| Al–Si (LM 6) | Si 12.0 | Sand cast (modified) | 20 | 51 | 50 | Rotating beam, 24 h at temp. |
| | | | 100 | 43 | | |
| | | | 200 | 35 | | |
| | | | 300 | 25 | | |
| Al–Cu (2219) | Cu 6.0 | Forged T6 | 20 | 117 | 120 | Reverse bending stresses |
| | | | 150 | 65 | | |
| | | | 200 | 62 | | |
| | | | 250 | 46 | | |
| | | | 300 | 39 | | |
| | | | 350 | 23 | | |

**Table 3.6** *(continued)*

| Material (specification) | Nominal composition % | Temp. Condition | (unnotched) °C | Endurance MPa | MHz | Remarks |
|---|---|---|---|---|---|---|
| Al–Si–Cu (LM 22) | Si 4.6 Cu 2.8 | Sand cast | 20 100 200 300 | 62 54 60 42 | 50 | Rotating beam |
| Al–Cu–Si–Mn (2014) | Cu 4.5 Si 0.8 Mn 0.8 | Forgings T6 | 148 203 260 | 65 45 25 | 100 | Rotating beam |
| Al–Cu–Mn–Mg (2014) | Cu 4.0 Mn 0.5 Mg 0.5 | Extruded T4 rod | 25 148 203 260 | 103 93 65 31 | 500 | Rotating beam, 100 days at temp. |
| Al–Cu–Mg–Si–Mn (2014) | Cu 4.4 Mg 0.7 Si 0.8 Mn 0.8 | Forgings T4 | 20 150 200 250 300 | 119 90 62 54 39 | 120 | Reversed bending |
| | | Forgings T6 | 20 150 200 250 300 | 130 79 57 39 39 | 120 | Reversed bending |
| Al–Cu–Mg–Ni (2218) | Cu 4.0 Mg 1.5 Ni 2.0 | Forged | 20 148 203 260 | 117 103 65 45 | 500 100 100 100 | Rotating beam after prolonged heating |
| | | Chill cast T6 | 20 100 200 300 | 100 105 108 80 | 50 | Rotating beam, 24 h at temp. |
| Al–Ni–Cu | Ni 2.5 Cu 2.2 | Forged T6 | 20 150 200 250 300 350 | 113 82 70 59 39 39 | 120 | Reversed bending |
| Al–Si–Cu–Mg–Ni (LM 13) | Si 12.0 Cu 1.0 Mg 1.0 | Chill cast (Lo-Ex) | 20 100 200 300 | 97 107 97 54 | 50 | Rotating beam, 24 h at temp. |
| Al–Zn–Mg–Cu (7075) | Zn 5.6 Mg 2.5 Cu 1.6 Cr 0.2 | Plate T6 | 24 149 204 260 | 151 83 59 48 | 500 | Reversed bending |

T4 = Solution treated and naturally aged, will respond to precipitation treatment.
T6 = Solution treated and artificially aged.

## 3.2   Mechanical properties of magnesium and magnesium alloys

**Table 3.7**   MAGNESIUM AND MAGNESIUM ALLOYS (WROUGHT) – TYPICAL MECHANICAL PROPERTIES AT ROOM TEMPERATURE

| Material | Nominal* composition % | Form | DTD or BS (Air) | BS (Gen. Eng.) | ASTM | Elektron | Tension Proof stress 0.2% MPa | UTS MPa | Elong. % | Compression Proof stress 0.2% MPa | Hardness VPN 30 kg |
|---|---|---|---|---|---|---|---|---|---|---|---|
| Mg | Mg 99.9 | Sheet, annealed | – | 3370-MAG-S-101M | – |  | 69 | 185 | 4 | – | 30–35 |
|  |  | Bar, extruded | – |  |  |  | 100 | 232 | 6 | – | 35–45 |
| Mg–Mn | Mn 1.5 | Sheet | 118C |  | – | AM503 | 100 | 232 | 6 | – | 35–45 |
|  |  | Extruded bar (1 in diam.) | 142B | 3373-MAG-E-101M | M1A-F, B107 |  | 162 | 263 | 7 | 124 | 45–55 |
|  |  | Extruded tube | 737A | 3373-MAG-E-101M | M1A, B107 |  | 154 | 247 | 6 | – | 45–55 |
| Mg–Al–Zn | Al 3.0 Zn 1.0 Mn 0.3 | Sheet, annealed |  | 3370-MAG-S-1110 | AZ31, B90 | AZ31 | 131 | 232 | 13 | – | 50–60 |
|  |  | half hard |  | 3370-MAG-S-111M | AZ31, B90 |  | 170 | 263 | 10 | 100 | 55–70 |
|  |  | Extruded bar and sections |  | 3373-MAG-E-111M | AZ31, B107 |  | 162 | 255 | 11 | 93 | 50–60 |
|  | Al 6.0 Zn 1.0 Mn 0.3 | Forgings | 2L513 | 3372-MAG-F-121M | AZ61, B91 | AZM | 183 | 293 | 8 | 147 | 60–70 |
|  |  | Extruded bar and sections | 2L512 | 3373-MAG-E-121N | AZ61, B107 |  | 183 | 293 | 8 | 147 | 55–70 |
|  |  | Extruded tube | 2L503 | 3373-MAG-E-121M | AZ61, B107 |  | 170 | 278 | 8 | 147 | 60–70 |
|  | Al 8.0 Zn 0.5 Mn 0.3 | Forgings | 88C | – | AZ80A, B91 | AZ855 | 208 | 293 | 8 | 185 | 65–75 |
| Mg–Zn–Mn | Zn 2.0 Mn 1.0 | Sheet, annealed | 5091 | 3370-MAG-S-1310 |  | ZM21 | 131 | 232 | 13 |  |  |
|  |  | half breed | 5101 | 3370-MAG-S-131M |  |  | 170 | 263 | 10 |  |  |
|  |  | Extruded bar sections | – | 3373-MAG-E-131M |  |  | 162 | 255 | 11 |  |  |
| Mg–Zn–Zr | Zn 1.0 Zr 0.6 | Sheet | 2L514 | 3370-MAG-S-141M | – | ZW1 | 178 | 263 | 10 | 154 | 55–70 |
|  |  | Extruded bar and sections | 2L508 | 3373-MAG-E-141M | – |  | 208 | 293 | 13 | 177 | 60–75 |
|  |  | Extruded tube | 2L509 | 3373-MAG-E-141M | – |  | 193 | 278 | 7 | – | 60–75 |
|  | Zn 3.0 | Sheet | 2L504 | 3370-MAG-S-151M | – | ZW3 | 185 | 270 | 8 | 154 | 60–70 |

|  | Composition | Form | Spec. | Spec. | Spec. | Symbol | 0.2% PS | TS | Elong. % | | Range |
|---|---|---|---|---|---|---|---|---|---|---|---|
|  | Zr 0.6 | Forgings | 2L514 | 3372-NAG-F-151M | – | – | 224 | 309 | 8 | 193 | 60–80 |
|  |  | Extruded bar and sections (1 in diam.) | 2L505 | 3373-MAG-E-151M | – | – | 239 | 309 | 18 | 213 | 65–75 |
|  | Zn 5.5 Zr 0.6 | Bars and sections Heat treated | 5041A | 3373-MAG-E-161TE | ZK60A-T5, B107-70 | ZW6 | 270 | 340 | 10 | 255 | 60–80 |
| Mg–Zn–Cu–Mn | Zn 6.5 Cu 1.3 Mn 0.8 | Bars and sections Heat treated | – | – | ZC71-T6, B107 | ZC71 | 340 | 360 | 6 | – | – |
| Mg–Th–Zn–Zr** | Th 0.8 | Extruded bar and sections 5111 | – | – | – | ZTy | 147 | 263 | 18 | – | 50–70 |
| (Creep resistant) | Zn 0.5 Zr 0.6 | Forgings 5111 | – | – | – |  | 147 | 232 | 13 | – | 50–70 |
| Mg–Th–Mn** (Creep resistant) | Th 2.0 Mn 0.75 | Sheet | – | HM21-T8, B90 | – | – | 165 | 247 | 9 | 179 | – |
|  | Th 3.0 Mn 1.2 | Extruded bar and sections | – | HM31-T5 | – | – | 227 | 287 | 8 | 185 | – |

Nuclear alloys: Two wrought magnesium alloys (Magnox AL80; Mg0.75Al-0.005 Be and MN70; Mg0.75 Mn) of interest only for their nuclear and high-temperature tensile properties similar to those of AM503.

*It is usual to add 0.2–0.4% Mn to alloys containing aluminium to improve corrosion resistance. M = As manufactured. O = Fully annealed. TE = Precipitation treated.

**Thorium-containing alloys are being replaced by alternative Mg alloys.

Table 3.8   MAGNESIUM AND MAGNESIUM ALLOYS (CAST) TYPICAL MECHANICAL PROPERTIES AT ROOM TEMPERATURE

| Material | Nominal* composition % | Condition | DTD or BS (Air) | BS (Gen. Eng.) | ASTM | Elektron | Tension Proof stress 0.2% MPa | UTS MPa | Elong. % | Compression Proof stress 0.2% MPa | Brinell hardness‡ VPN 30 kg |
|---|---|---|---|---|---|---|---|---|---|---|---|
| Mg–Zr | Zr 0.6 | AC | – | – | KIA, B80 | ZA | 51 | 185 | 2.0 | 54 | 40–50 |
| Mg–Al–Zn | Al 6.0 Zn 3.0 | AC | – | – | AZ63A-F, B80 | – | 97 | 199 | 5 | 97 | 50 |
| | | TB | – | – | AZ63A-T4, B80 | | 97 | 275 | 10 | 97 | 55 |
| | | TF | – | – | AZ63A-T6, B80 | | 131 | 275 | 5 | 131 | 73 |
| | Al 8.0 Zn 0.4 | AC | – | 2970 MAG 1-M | | A8 | 86 | 158 | 4 | 86 | 50–60 |
| | | TB | 3L122 | 2970 MAG 1-TB | AZ81A-T4, B80 | | 82 | 247 | 11 | 82 | 50–60 |
| | Al 9.5 Zn 0.4 | AC | – | 2970 MAG 3-M | AZ91C-F, B80 | AZ91 | 93 | 154 | 2 | 93 | 55–65 |
| | | TB | 3L124 | 2970 MAG 3-TB | AZ1C-T4, B80 | | 90 | 232 | 6 | 90 | 55–65 |
| | | TF | 3L125 | 2970 MAG 3-TF | AZ91C-T6, B80 | | 127 | 239 | 2 | 124 | 75–85 |
| | | Die cast | – | – | AZ91B-F, B94 | | 111 | 216 | 3 | 108 | 60–70 |
| | Al 9.0 Zn 2.0 | AC | – | – | AZ92A, B80 | – | 97 | 165 | 2 | 97 | 65 |
| | | TB | – | – | – | | 97 | 275 | 8 | 97 | 63 |
| | | TF | – | – | – | | 145 | 275 | 2 | 145 | 84 |
| Mg–Zn–Zr | Zn 4.5 Zr 0.7 | TE | 2L127 | 2970 MAG 4-TE | ZK51A-T5, B80 | Z5Z | 161 | 263 | 6 | 162 | 65–75 |
| Mg–Zn–RE–Zr | Zn 4.0 RE 1.2 Zr 0.7 | TE | 2L128 | 2970 MAG 5-TE | ZE41A-T5, B80 | RZ5 | 150 | 216 | 5 | 139 | 55–75 |
| | Zn 6.0 RE 2.5 Zr 0.7 | TF§ | 5045 | – | | ZE63 | 190 | 295 | 7 | 190 | 70–80 |
| Mg–RE–Zn–Zr (Creep resistant to 250°C) | RE 2.7 Zn 2.2 Zr 0.7 | TE | 2L126 | 2970 MAG 6-TE | EZ33A-T5, B80 | ZRE1 | 95 | 162 | 4.5 | 93 | 50–60 |
| Mg–Th–Zn–Zr** (Creep resistant to 350°C) | Th 3.0 Zn 2.2 Zr 0.7 | TE | 5005A | 2970 MAG 8-TE | HZ32A-T5, B80 | ZT1 | 93 | 216 | 7 | 93 | 50–60 |
| Mg–Zn–Th–Zr** | Zn 5.5 Th 1.8 Zr 0.7 | TE | 5015A | 2970 MAG 9-TE | ZH62A-T5, B80 | TZ6 | 167 | 270 | 8 | 162 | 65–75 |

| Alloy | Composition | Condition | | | | | | | | | |
|---|---|---|---|---|---|---|---|---|---|---|---|
| Mg–Th–Zr** | Th 3.0 Zr 0.7 | TF | – | – | HK31A-T6, B80 | MTZ | 93 | 208 | 5 | 93 | 50–60 |
| Mg–Ag–RE‡–Zr | Ag 2.5 RE 2.0‡ Zr 0.6 | TF | 5025A 5035A | 2970 MAG 12-TF | | MSR-A MSR-B | 187 204 | 247 260 | 5 3 | 178 193 | 65–80 65–80 |
| | Ag 2.5 RE 2.0‡ Zr 0.6 | TF | 5055 | – | QE22A-T6, B80 | QE22 | 200 | 260 | 4 | 195 | 65–80 |
| Mg–RE(D)–Ag–Zr–Cu | RE(D)2.2 Ag 1.5 Zr 0.6 Cu 0.07 | TF | 5055 | 2970 MAG 13-TF | EQ21A-T6, B80 | EQ21 | 195 | 261 | 4 | – | 75–90 |
| Mg–Ag–Th–RE‡Zr** | Ag 2.5 RE 1.0‡ Th 1.0 | TF | – | – | QH21A-T6, B80 | QH21A | 210 | 270 | 4 | 200 | 65–80 |
| Mg–Y–RE(Δ)–Zr | Y 4.0 Zr 0.7 RE(Δ)3.4 Zr 0.6 | TF | – | – | WE43-T6, B80 | WE43 | 185 | 265 | 7 | – | 75–90 |
| | Y 5.1 RE(Δ)3.0 Zr 0.6 | TF | – | 2970 MAG 14-TF | WE54-T6, B80 | WE54 | 205 | 280 | 4 | – | 75–90 |
| Mg–Zn–Cu–Mn | Zn 6.0 Cu 2.7 Mn 0.5 | TF | – | – | ZC63-T6, B80 | ZC63 | 158 | 242 | 4.5 | – | 55–65 |

*It is usual to add 0.2–0.4% Mn to alloys containing aluminium to improve corrosion resistance. RE = Cerium mischmetal containing approx. 50% cerium. RE(Δ) = Neodymium plus Heavy Rare Earth metals. RE(D) = Neodymium enriched mischmetal.

†Brinell tests with 500 kg on 10 mm ball for 30 s.

‡Fractionated rare earth metals: MSR-A contains 1.7%; MSR-B contains 2.5%.

§Solution heat treated in an atmosphere of hydrogen.

AC = Sand cast. TE = Precipitation heat treated.

TB = Solution heat treated. TF = Fully heat treated.

**Thorium-containing alloys are being replaced by alternative Mg alloys.

**Table 3.9** MAGNESIUM AND MAGNESIUM ALLOYS (excluding high temperature alloys for which see table 3.10) – TYPICAL TENSILE PROPERTIES AT ELEVATED TEMPERATURES

| Material | Nominal composition* % | Form and condition | Test temp. °C | *'Short-time' tension*[†] | | | |
|---|---|---|---|---|---|---|---|
| | | | | Young's modulus GPa | 0.2% proof stress MPa | UTS MPa | Elong. % |
| Mg | Mg 99.95 | Forged | 20 | 45 | – | 170 | 5 |
| | | | 100 | – | – | 128 | 8 |
| | | | 150 | – | – | 93 | 16 |
| | | | 200 | – | – | 54 | 43 |
| Mg–Al–Zn | Al 8.0 Zn 0.4 (A8) | Sand cast | 20 | 45 | 86 | 158 | 4 |
| | | | 100 | 34 | 76 | 154 | 5 |
| | | | 150 | 32 | 65 | 145 | 11 |
| | | | 200 | 25 | 62 | 100 | 20 |
| | | | 250 | – | – | 75 | 27 |
| | | Sand cast and solution treated | 20 | 45 | 82 | 247 | 11 |
| | | | 100 | 34 | 73 | 202 | 16 |
| | | | 150 | 33 | 65 | 154 | 21 |
| | | | 200 | 28 | 62 | 116 | 25 |
| | | | 250 | – | – | 85 | 21 |
| | (AZ855) | Forged | 20 | 45 | 221 | 309 | 8 |
| | | | 150 | – | 153 | 216 | 25 |
| | | | 200 | – | 102 | 154 | 28 |
| | Al 9.5 Zn 0.4 (AZ91) | Sand cast | 20 | 45 | 93 | 154 | 2 |
| | | | 100 | – | – | 131 | 2 |
| | | | 150 | – | – | 122 | 6 |
| | | | 200 | – | – | 108 | 25 |
| | | | 250 | – | – | 77 | 34 |
| | | Sand cast and solution treated | 20 | 45 | 90 | 232 | 6 |
| | | | 100 | – | – | 222 | 12 |
| | | | 150 | – | – | 196 | 16 |
| | | | 200 | – | – | 139 | 20 |
| | | Sand cast and fully heat treated | 20 | 45 | 127 | 239 | 2 |
| | | | 100 | 40 | 91 | 232 | 6 |
| | | | 150 | 37 | 77 | 185 | 25 |
| | | | 200 | 28 | 62 | 133 | 34 |
| | | | 250 | 19 | 46 | 103 | 30 |
| Mg–Zn–Zr | Zn 4.5 Zr 0.7 (Z5Z) | Sand cast and heat treated | 20 | 45 | 161 | 263 | 6 |
| | | | 100 | 34 | 124 | 185 | 14 |
| | | | 150 | 28 | 102 | 145 | 20 |
| | | | 200 | 22 | 79 | 113 | 23 |
| | | | 250 | 19 | 57 | 85 | 20 |
| | Zn 3.0 Zr 0.6 (ZW3) | Extruded | 20 | 45 | 255 | 309 | 18 |
| | | | 100 | 40 | 162 | 182 | 33 |
| | | | 200 | 22 | 46 | 127 | 56 |
| | | | 250 | 12 | 11 | 100 | 71 |
| | | Sheet | 20 | 45 | 195 | 270 | 10 |
| | | | 100 | 40 | 120 | 165 | 33 |
| | | | 150 | 33 | 74 | 116 | 42 |
| | | | 200 | – | – | 76 | 51 |
| | | | 250 | – | – | 49 | 59 |
| Mg–Zn–RE–Zr | Zn 4.0 Re 1.2 Zr 0.7 (RZ5) | Sand cast Sand cast treated | 20 | 45 | 150 | 216 | 4 |
| | | | 20 | 41 | 134 | 195 | 6 |
| | | | 150 | 40 | 120 | 167 | 19 |
| | | | 200 | 38 | 99 | 131 | 29 |
| | | | 250 | 33 | 74 | 99 | 35 |

**Table 3.9** *(continued)*

| Material | Nominal composition* % | Form and condition | Test temp. °C | Young's modulus GPa | 0.2% proof stress MPa | UTS MPa | Elong. % |
|---|---|---|---|---|---|---|---|
| | | | | | 'Short-time' tension[†] | | |
| Mg–Zn–Th–Zr** | Zn 5.5 | Sand cast | 20 | 45 | 161 | 270 | 9 |
| | Th 1.8 | and heat | 100 | 34 | 134 | 224 | 22 |
| | Zr 0.7 | treated | 150 | 31 | 110 | 178 | 26 |
| | (TZ 6) | | 200 | 28 | 82 | 130 | 26 |
| | | | 250 | 26 | 52 | 91 | 25 |
| Mg–Ag– | Ag 2.5 | Sand cast | 20 | 45 | 201 | 259 | 4 |
| RE–Zr | RE(D)2.0 | and fully | 100 | 41 | 185 | 232 | 12 |
| (D) | Zr 0.6 | heat | 150 | 40 | 171 | 210 | 16 |
| | (QE22) | treated | 200 | 38 | 154 | 185 | 20 |
| | | | 250 | 34 | 102 | 142 | 27 |
| | | | 300 | 31 | 68 | 88 | 59 |
| Mg–RE(D)– | Re(D)2.2 | Sand cast | 20 | 45 | 195 | 261 | 4 |
| Ag–Zr–Cu | Ag 1.5 | and fully | 100 | 43 | 189 | 230 | 10 |
| | Zr 0.6 | heat | 150 | 42 | 180 | 211 | 16 |
| | Cu 0.07 | treated | 200 | 41 | 170 | 191 | 16 |
| | (EQ21) | | 250 | 39 | 152 | 169 | 15 |
| | | | 300 | 35 | 117 | 132 | 10 |
| Mg–Ag–Re(D)** | Ag 2.6 | Sand cast | 20 | 45 | 210 | 270 | 4 |
| Th–Zr | RE(D)1.0 | and fully | 100 | 41 | 199 | 242 | 17 |
| ‡ | Th 1.0 | heat | 150 | 40 | 190 | 224 | 20 |
| | Zr 0.6 | treated | 200 | 38 | 183 | 205 | 18 |
| | (OH21) | | 250 | 37 | 167 | 185 | 19 |
| | | | 300 | 33 | 120 | 131 | 20 |
| Mg–Y–RE(Δ)–Zr | Y 4.0 | Sand cast | 20 | 45 | 185 | 265 | 7 |
| | RE(Δ)3.4 | and fully | 150 | 42 | 175 | 250 | 6 |
| | Zr 0.6 | heat | 200 | 39 | 170 | 245 | 11 |
| | (WE43) | treated | 250 | 37 | 160 | 220 | 18 |
| | | | 300 | 35 | 120 | 160 | 40 |
| | Y 5.1 | Sand cast | 20 | 45 | 205 | 280 | 4 |
| | RE(Δ)3.0 | and fully | 100 | 43 | 197 | 260 | 4.5 |
| | Zr 0.6 | heat | 150 | 42 | 195 | 255 | 5 |
| | (WE54) | treated | 200 | 41 | 183 | 241 | 6.5 |
| | | | 250 | 39 | 175 | 230 | 9 |
| | | | 300 | 36 | 117 | 184 | 14.5 |
| Mg–Zn–Cu–Mn | Zn 6.0 | Sand cast | 20 | 45 | 158 | 242 | 4.5 |
| | Cu 2.7 | and fully | 100 | – | 141 | 215 | 9 |
| | Mn 0.5 | heat | 150 | – | 134 | 179 | 14 |
| | (Zc63) | treated | 200 | – | 118 | 142 | 11 |
| | Zn 6.5 | Extruded | 20 | 45 | 325 | 350 | 6 |
| | Cu 1.3 | and fully | 100 | 40 | 206 | 259 | 16 |
| | Mn 0.8 | heat | 200 | 32 | 115 | 163 | 14 |
| | (Zc71) | treated | | | | | |

*It is usual to add 0.2–0.4% Mn to alloys containing aluminium to improve corrosion resistance.

[†]In accordance with BS1094: 1943; 1 h at temperature and strain rate 0.1–0.25 in in$^{-1}$ min$^{-1}$.

‡Tested according to BS4A4. RE = Cerium mischmetal containing approx. 50% Ce. RE(D) = Neodymium enriched mischmetal. RE(Δ) = Neodymium plus Heavy Rare Earth metals.

**Thorium-containing alloys are being replaced by alternative Mg alloys.

**Table 3.10**    HIGH TEMPERATURE MAGNESIUM ALLOYS – TENSILE PROPERTIES AT ELEVATED TEMPERATURE

| Material | Nominal composition* % | Form and condition | Test temp. °C | *'Short-time' tension*[†] | | | |
|---|---|---|---|---|---|---|---|
| | | | | *Young's modulus* GPa | *0.2% proof stress* MPa | *UTS* MPa | *Elong.* % |
| Mg–RE–Zn | RE 2.7 Zn 2.2 Zr 0.7 (ZRE1) | Sand cast and heat treated | 20 | 45 | 93 | 162 | 4.5 |
| | | | 100 | 40 | 79 | 150 | 11 |
| | | | 150 | 38 | 76 | 139 | 19 |
| | | | 200 | 36 | 74 | 125 | 26 |
| | | | 250 | 33 | 65 | 107 | 35 |
| | | | 300 | 28 | 48 | 85 | 51 |
| | | | 350 | 21 | 26 | 56 | 90 |
| Mg–Th–Zr** | Th 3.0 Zr 0.7 (HK31) (MTZ) | Sand cast and fully heat treated | 20 | 45 | 93 | 208 | 4 |
| | | | 100 | 40 | 88 | 188 | 10 |
| | | | 150 | 38 | 86 | 174 | 13 |
| | | | 200 | 38 | 85 | 162 | 17 |
| | | | 250 | 36 | 83 | 150 | 20 |
| | | | 300 | 34 | 73 | 136 | 22 |
| | | | 350 | 29 | 56 | 103 | 23 |
| Mg–Th–Zn–Zr** | Th 3.0 Zn 2.2 Zr 0.7 (ZT1) | Sand cast . and heat treated | 20 | 45 | 93 | 216 | 9 |
| | | | 100 | 36 | 88 | 159 | 23 |
| | | | 150 | 34 | 79 | 131 | 27 |
| | | | 200 | 33 | 65 | 108 | 33 |
| | | | 250 | 33 | 56 | 90 | 38 |
| | | | 300 | 31 | 49 | 76 | 41 |
| | | | 350 | 28 | 45 | 63 | 34 |
| | Th 0.8 Zn 0.5 Zr 0.6 (ZTY) | Sheet | 20 | 45 | 181 | 266 | 10 |
| | | | 100 | 41 | 179 | 224 | 10 |
| | | | 150 | 41 | 176 | 201 | 11 |
| | | | 200 | 40 | 165 | 171 | 15 |
| | | | 250 | 40 | 124 | 134 | 20 |
| | | | 300 | 34 | 73 | 96 | 27 |
| | | | 350 | 29 | 17 | 56 | 38 |
| Mg–Ag– RE(D)–Zr | Ag 2.5 RE(D)2.0 Zr 0.6 (QE22) | Sand cast and fully heat treated | | | | | |
| Mg–RE(D)– Ag–Zr–Cu | RE(D)2.2 Ag 1.5 Zr 0.6 Cu 0.07 (EQ21) | Sand cast and fully heat treated | | | | | |
| Mg–Ag–RE(D)– Th–Zr** | Ag 2.5 RE(D)1.0 Th 1.0 Zr 0.6 (QH21) | Sand cast and fully heat treated | | High strength cast alloys with good elevated temperature properties – for which *see* Table 3.9 | | | |
| Mg–Y–RE(Δ)–Zr | Y 4.0 RE(Δ)3.4 (WE43) | Sand cast and fully treated | | | | | |
| | Y 5.1 RE(Δ)3.0 Zr 0.6 (WE54) | Sand cast and fully eat treated | | | | | |

*It is usual to add 0.2–0.4% Mn to alloys containing aluminium to improve corrosion resistance.
[†]In accordance with BS 1094: 1943; 1 h at temperature; strain rate 0.1–0.25 in$^{-1}$ min$^{-1}$.
RE = Cerium mischmetal containing approx. 50% Ce. RE(D) = neodymium-enriched mischmetal.
RE(Δ) = Neodymium plus Heavy Rare Earths.
**Thorium containing alloys are being replaced by alternative Mg alloys.

**Table 3.11** HIGH-TEMPERATURE MAGNESIUM ALLOYS – LONG-TERM CREEP RESISTANCE

| Material | Nominal composition % | Form and Condition | Temp. °C | Time† h | Stress to produce specified creep strains% | | | | |
|---|---|---|---|---|---|---|---|---|---|
| | | | | | 0.05 MPa | 0.1 MPa | 0.2 MPa | 0.5 MPa | 1.0 MPa |
| Mg–RE–Zn–Zr | RE 2.7<br>Zn 2.2<br>Zr 0.7<br>(ZRE1) | Sand cast and heat treated | 200 | 100<br>500<br>1 000 | 52<br>41<br>36 | 66<br>54<br>47 | 71<br>65<br>58 | –<br>–<br>– | –<br>–<br>– |
| | | | 250 | 100<br>500<br>1 000 | 23<br>11<br>– | 28<br>19<br>14 | 32<br>24<br>20 | 36<br>30<br>26 | –<br>34<br>30 |
| | | | 315 | 100<br>500<br>1 000 | 5.6<br>–<br>– | 7.4<br>5.2<br>4.3 | 8<br>6.5<br>5.6 | –<br>–<br>– | –<br>–<br>– |
| | Zn 4.0<br>RE 1.2<br>Zr 0.7<br>(RZ5) | Sand cast and heat treated | 100 | 100<br>500<br>1 000 | –<br>–<br>– | 97<br>–<br>– | 111<br>106<br>103 | 117<br>117<br>116 | –<br>–<br>– |
| | | | 150 | 100<br>500<br>1 000 | 77<br>–<br>– | 86<br>75<br>70 | 97<br>88<br>83 | 101<br>96<br>91 | 107<br>100<br>97 |
| | | | 200 | 100<br>500<br>1 000 | 29<br>22<br>20 | 43<br>28<br>23 | 52<br>37<br>31 | 67<br>52<br>43 | 73<br>64<br>53 |
| | | | 250 | 100<br>500<br>1 000 | 6.2<br>4.3<br>3.9 | 12<br>6.2<br>5.4 | 19<br>8.6<br>6.9 | 32<br>15<br>12 | 39<br>19<br>15 |
| Mg–Th–Zr** | Th 3.0<br>Zr 0.7<br>(HK31)<br>(MTZ) | Sand cast and fully heat treated | 200 | 100<br>1 100 | 31*<br>– | 45*<br>– | 63*<br>62* | 97*<br>100* | 111*<br>– |
| | | | 260 | 100<br>1 000 | –<br>– | 28*<br>– | 43*<br>29* | 65*<br>45* | –<br>– |
| | | | 315 | 100 | 9.3* | 14* | 19* | 27* | 32* |
| Mg–Th–Zn–Zr** | Th 0.8<br>Zn 0.5<br>Zr 0.6<br>(ZTY) | Sheet | 250 | 100 | Stress of 46 MPa (3 tonf in⁻²) produced 0.03% creep strain Stress of 46 MPa (3 tonf in⁻²) produced 0.03% creep strain | | | | |
| Mg–Th–Zn–Zr* | Th 3.0<br>Zn 2.2<br>Zr 0.7<br>(ZT1) | Sand cast and heat treated | 250 | 100<br>500<br>1 000 | 42<br>35<br>31 | 50<br>43<br>39 | 56<br>51<br>48 | 63<br>58<br>56 | 66<br>63<br>61 |
| | | | 300 | 100<br>500<br>1 000 | 23<br>19<br>17 | 28<br>21<br>19 | 35<br>25<br>21 | 46<br>36<br>32 | 52<br>41<br>36 |
| | | | 325 | 100<br>500<br>1 000 | 14<br>12<br>10 | 19<br>13<br>12 | 24<br>16<br>13 | 29<br>21<br>15 | 36<br>25<br>20 |
| | | | 350 | 100<br>500<br>1 000 | 10<br>–<br>– | 12<br>9<br>8 | 18<br>10<br>8 | 21<br>12<br>9 | 23<br>14<br>10 |
| | | | 375 | 100<br>500<br>1 000 | –<br>–<br>– | 8<br>–<br>– | 11<br>–<br>– | 12<br>8<br>– | 13<br>9<br>8 |
| | Zn 5.5<br>Th 1.8<br>Zr 0.7<br>(TZ6) | Sand cast and heat treated | 150 | 100<br>500<br>1 000 | 51<br>36<br>26 | 66<br>56<br>51 | 82<br>69<br>63 | 96<br>85<br>80 | 102<br>94<br>90 |
| | | | 200 | 100<br>500<br>1 000 | 26<br>15<br>11 | 32<br>22<br>17 | 45<br>26<br>20 | 56<br>40<br>31 | 62<br>49<br>40 |

**Table 3.11**    (*continued*)

| Material | Nominal composition % | | Form and Condition | Temp. °C | Time† h | Stress to produce specified creep strains% | | | | |
|---|---|---|---|---|---|---|---|---|---|---|
| | | | | | | 0.05 MPa | 0.1 MPa | 0.2 MPa | 0.5 MPa | 1.0 MPa |
| Mg–Ag– RE(D)–Zr | Ag RE(D) Zr (QE22) | 2.5 2.0 0.6 | Sand cast and fully heat treated | 200 | 100 500 1 000 | 55 – – | 74 54 46 | 88 65 56 | – 82 73 | – 89 79 |
| | | | | 250 | 100 500 1 000 | 18 – – | 26 15 10 | 33 22 16 | – 28 22 | – 31 26 |
| Mg–RE(D)–Ag– Zr–Cu | RE(D) Ag Zr Cu (EQ21) | 2.2 1.5 0.6 0.07 | Sand cast and fully heat treated | 200 | 100 500 1 000 | – – – | 78 57 48 | 95 71 62 | 116 88 76 | – – – |
| | | | | 250 | 100 500 1 000 | – – – | 29 18 14 | 36 22 19 | 42 30 24 | – – – |
| Mg–Ag–RE(D)– Th–Zr** | Ag RE(D) Th Zr (QH21) | 2.5 1.0 1.0 0.6 | Sand cast and fully heat treated | 250 | 100 500 1 000 | 22 – – | 32 20 – | 39 26 21 | – 32 26 | – 36 30 |
| Mg–Y–RE(Δ)–Zr | Y RE(Δ) Zr (WE43) | 4.0 3.4 0.6 | Sand cast and fully heat treated | 200 | 100 500 1 000 | – – – | 148 – – | 161 115 96 | 173 148 139 | – – – |
| | | | | 250 | 100 500 1 000 | – – – | 44 – – | 61 46 39 | – – – | – – – |
| | Y RE(Δ) Zr (WE54) | 5.1 3.0 0.6 | Sand cast and fully heat treated | 200 | 100 500 1 000 | – – – | 160 120 120 | 165 140 132 | – – – | – – – |
| | | | | 250 | 100 500 1 000 | – – – | 47 43 16 | 61 40 32 | 81 58 48 | – – – |
| Mg–Zn–Cu–Mn | Zn Cu Mn (ZC63) | 6.0 2.7 0.5 | Sand cast and fully heat treated | 150 | 100 500 1 000 | – – – | 94 82 74 | 99 92 89 | 104 98 95 | – – – |
| | | | | 200 | 100 500 1 000 | – – – | 60 51 42 | 63 55 49 | 67 61 55 | – – – |

*Total strains.
†4–6 h heating to test temperature followed by 16 h soaking at test temperature.
RE = Cerium mischmetal containing approx. 50% Ce.
RE(D) = Neodymium-enriched mischmetal.
RE(Δ) = Neodymium plus Heavy Rare Earth metals.
**Thorium-containing alloys are being replaced by alternative Mg alloys.

**Table 3.12** HIGH-TEMPERATURE MAGNESIUM ALLOYS – SHORT-TERM CREEP RESISTANCE

| Material | Nominal composition % | | Form and condition | Temp. °C | Time† s | *Stress to produce specified creep strains%* | | | | | Stress to fracture MPa |
|---|---|---|---|---|---|---|---|---|---|---|---|
| | | | | | | 0.05 MPa | 1.0 MPa | 2.0 MPa | 5.0 MPa | 10.0 MPa | |
| Mg–RE–Zn–Zr | Re | 2.7 | Sand cast and heat treated | 200 | 30 | – | – | 98 | 118 | 130 | 136 |
| | Zn | 2.2 | | | 60 | – | – | 97 | 117 | 128 | 134 |
| | Zr | 0.7 | | | 600 | – | – | 96 | 116 | 125 | 129 |
| | (ZRE1) | | | 250 | 30 | 76 | 84 | 92 | 111 | 123 | 130 |
| | | | | | 60 | 74 | 83 | 91 | 110 | 120 | 129 |
| | | | | | 600 | 73 | 82 | 89 | 108 | 114 | 125 |
| | | | | 315 | 30 | 52 | 59 | 73 | 80 | 85 | 90 |
| | | | | | 60 | 51 | 58 | 69 | 76 | 83 | 88 |
| | | | | | 600 | 42 | 49 | 56 | 62 | 68 | 73 |
| | Zn | 4.0 | Sand cast and heat treated | 200 | 30 | 100 | 107 | 116 | 127 | – | 136 |
| | RE | 1.2 | | | 60 | 99 | 105 | 114 | 124 | – | 134 |
| | Zr | 0.7 | | | 600 | 86 | 99 | 103 | 114 | – | 125 |
| | (RZ5) | | | 250 | 30 | 86 | 90 | 94 | 99 | – | 116 |
| | | | | | 60 | 83 | 88 | 91 | 96 | – | 113 |
| | | | | | 600 | 71 | 76 | 81 | 86 | – | 93 |
| | | | | 315 | 30 | 62 | 69 | 76 | 79 | 83 | 86 |
| | | | | | 60 | 59 | 66 | 73 | 76 | 79 | 82 |
| | | | | | 600 | 48 | 53 | 59 | 64 | 67 | 69 |
| Mg–Th–Zr* | Th | 3.0 | Sand cast and fully heat treated | 250 | 30 | 96 | 103 | 119 | 138 | – | 145 |
| | Zr | 0.7 | | | 60 | 95 | 103 | 118 | 137 | – | 145 |
| | (HK31) | | | | 600 | 94 | 102 | 117 | 137 | – | 144 |
| | (MTZ) | | | 315 | 30 | 80 | 88 | 103 | 117 | – | 128 |
| | | | | | 60 | 78 | 86 | 102 | 116 | – | 127 |
| | | | | | 600 | 74 | 82 | 96 | 107 | – | 120 |
| Mg–Th–Zn–Zr* | Th | 0.8 | Sheet | 250 | 30 | – | – | 110 | 159 | 163 | 165 |
| | Zn | 0.5 | | | 60 | – | – | 95 | 157 | 160 | 162 |
| | Zr | 0.6 | | | 600 | – | – | – | 145 | 149 | 151 |
| | (ZTY) | | | 350 | 30 | 20 | 32 | 48 | 80 | 93 | 102 |
| | | | | | 60 | 18 | 28 | 40 | 67 | 82 | 98 |
| | | | | | 600 | – | 15 | 20 | 31 | 42 | 66 |
| Mg–Th–Zn–Zr* | Th | 3.0 | Sand cast and heat treated | 200 | 30 | – | – | – | 100 | 118 | 125 |
| | Zn | 2.2 | | | 60 | – | – | – | 96 | 114 | 123 |
| | Zr | 0.7 | | | 600 | – | – | – | 85 | 103 | 114 |
| | (ZT1) | | | 250 | 30 | 58 | 65 | 71 | 84 | 102 | 111 |
| | | | | | 60 | 57 | 64 | 69 | 81 | 99 | 107 |
| | | | | | 600 | 56 | 63 | 68 | 74 | 86 | 98 |
| | | | | 315 | 30 | 55 | 60 | 64 | 73 | 76 | 82 |
| | | | | | 60 | 53 | 59 | 63 | 72 | 76 | 80 |
| | | | | | 600 | 50 | 59 | 61 | 71 | 74 | 77 |
| | Zn | 5.5 | Sand cast and heat treated | 200 | 30 | 96 | 113 | 120 | 128 | 137 | 144 |
| | Th | 1.8 | | | 60 | 93 | 109 | 117 | 124 | 133 | 137 |
| | Zr | 0.7 | | | 600 | 63 | 90 | 102 | 110 | 114 | 119 |
| | (TZ6) | | | 250 | 30 | 70 | 77 | 85 | 96 | 99 | 107 |
| | | | | | 60 | 65 | 74 | 80 | 90 | 94 | 99 |
| | | | | | 600 | 56 | 60 | 66 | 74 | 77 | 82 |
| | | | | 315 | 30 | 54 | 59 | 64 | 70 | 74 | 76 |
| | | | | | 60 | 52 | 57 | 62 | 66 | 70 | 73 |
| | | | | | 600 | 44 | 49 | 53 | 56 | 58 | 59 |

† 1 h heating to test temperature followed by 1 h soaking at test temperature.
RE = cerium mischmetal containing approx. 50% Ce.
*Therium-containing alloys are being replaced by alternative Mg alloys.

**Table 3.13**   MAGNESIUM AND MAGNESIUM ALLOYS – FATIGUE AND IMPACT STRENGTHS

| Material | Nominal* composition % | Condition | State ‡ | temp. °C | Fatigue strength† at specified cycles | | | | | | Test temp. °C | Impact strength§ for single blow fracture | |
|---|---|---|---|---|---|---|---|---|---|---|---|---|---|
| | | | | | $10^5$ MPa | $5 \times 10^5$ MPa | $10^6$ Mpa | $5 \times 10^6$ MPa | $10^7$ MPa | $5 \times 10^7$ MPa | | Unnotched J | Notched J |
| Mg–Mn | Mn 1.5 (AM503) | Extruded | U | 20 | 107 | 90 | 88 | 86 | 85 | 83 | 20 | 12–14 | 4–4.5 |
| | | | N | | 76 | 90 | 54 | 51 | 50 | 48 | | | |
| Mg–Al–Zn | Al 6.0 Zn 1.0 (AZM) | Extruded | U | 20 | 161 | 139 | 133 | 125 | 124 | 120 | 20 | 34–43 | 7–9.5 |
| | | | N | | 127 | 110 | 103 | 97 | 94 | 91 | | | |
| | Al 8 Zn 0.4 (A8) | Sand cast | U | 20 | 108 | 93 | 90 | 88 | 88 | 86 | 20 | 3–5 | 1.5–2 |
| | | | N | | 107 | 80 | 73 | 66 | 65 | 63 | | | |
| | | Sand cast and solution treated | U | 20 | 124 | 102 | 97 | 91 | 90 | 90 | 20 | 18–27 | 4.5–7 |
| | | | N | | 108 | 86 | 82 | 74 | 73 | 69 | | | |
| | | | U | 150 | 93 | 69 | 66 | 59 | 57 | 57 | | | |
| | | | U | 200 | 71 | 52 | 48 | 38 | 36 | 31 | | | |
| | Al 9.5 Zn 0.4 (AZ91) | Sand cast | U | 20 | 114 | 91 | 89 | 88 | 86 | 85 | −196 / 20 | 1.5 / 1.5–2.0 | 1–1.5 |
| | | | N | | 110 | 83 | 74 | 68 | 66 | 63 | | | |
| | | Sand cast and solution treated | U | 20 | 124 | 93 | 93 | 93 | 93 | – | 20 | 7–9.5 | 3–4 |
| | | | N | | 103 | 82 | 80 | 79 | 79 | 77 | | | |
| | | Sand cast and fully heat treated | U | 20 | 117 | 90 | 80 | 79 | 77 | 76 | 20 | 3–4 | 1–1.5 |
| | | | N | | 93 | 66 | 66 | 65 | 65 | 65 | | | |
| Mg–Zn–Zr | Zn 3.0 Zr 0.6 (ZW3) | Extruded | U | 20 | 151 | 137 | 134 | 128 | 127 | 124 | 20 | 23–31 | 9.5–12 |
| | | | N | | 124 | 99 | 93 | 91 | 90 | 88 | | | |
| | Zn 4.5 Zr 0.7 (Z5Z) | Sand cast and heat treated | U | 20 | 111 | 86 | 85 | 82 | 80 | 77 | 20 / −196 | 7–12 / 0.8 | 3–4 |
| | | | N | | 90 | 86 | 85 | 82 | 80 | 77 | | | |
| Mg–Zn–RE–Zr | Zn 4.0 RE 1.2 Zr 0.7 (RZ5) | Sand cast and heat treated | U | 20 | 124 | 99 | 97 | 97 | 96 | 94 | 20 / −196 | 4–5.5 / 0.7 | 1–2 |
| | | | N | | 108 | 93 | 91 | 88 | 86 | 83 | | | |
| | | | U | 150 | 97 | 85 | 80 | 73 | 69 | 65 | | | |
| | | | U | 200 | 93 | 74 | 69 | 62 | 59 | 54 | | | |
| | Zn 2.2 RE 2.7 | Sand cast and heat treated | U | 20 | 100 | 82 | 80 | 79 | 77 | 74 | 20 | 6–7.5 | 1–2 |
| | | | N | | 77 | 59 | 54 | 52 | 52 | 51 | | | |

| Alloy system | Composition | Condition | Code | Temp (°C) | 20 | 100 | 150 | 200 | 250 | 300 | Test temp (°C) | Value | Elong (%) | Range |
|---|---|---|---|---|---|---|---|---|---|---|---|---|---|---|
| | Zr (ZRE1) 0.7 | | U U U U | 150 200 250 300 | 69 | 60 | 59 | 57 | 57 | 57 | | | | 2.3–2.7 |
| | | | | | 68 | 59 | 56 | 52 | 51 | 51 | | | | |
| | | | | | 59 | 48 | 45 | 43 | 43 | 42 | | | | |
| | | | | | 49 | 39 | 37 | 37 | 36 | 34 | | | | |
| Mg–Ag–RE(D)–Zr | Zn 6, RE 2.5, Zr 0.6 (ZE63) | Sand cast and fully heat treated** | U N | 20 | 144 | 131 | 127 | 121 | 119 | 117 | –196 / 20 | 20 | 12.9–17.6 / 0.5 | |
| | | | | | 99 | 83 | 79 | 73 | 72 | 71 | | | | |
| Mg–Ag–RE(D)–Zr | Ag 2.5, RE(D) 2.5, Zr 0.6 (MSR–B) | Sand cast and fully heat treated | U Z N U | 20 / 200 250 | 119 | 103 | 103 | 103 | 102 | 100 | | | | |
| | | | | | 77 | 65 | 63 | 62 | 62 | 62 | | | | |
| | | | | | — | — | — | 90 | 88 | 86 | | | | |
| | | | | | — | 77 | 68 | 57 | 54 | 51 | | | | |
| Mg–Ag–RE(D)Th–Zr** | Ag 2.5, Re(D) 1.0, Th 1.0, Zr 0.6 (QH21) | Sand cast and fully heat treated | U Z U | 20 / 250 | 135 | 114 | 111 | 109 | 108 | 108 | | | | |
| | | | | | 86 | 72 | 69 | 64 | 63 | 62 | | | | |
| | | | | | 108 | 76 | 65 | 56 | 55 | 52 | | | | |
| Mg–Zn–Th–Zr** | Zn 5.5, Th 1.8, Zr 0.7 (TZ6) | Sand cast and heat treated | U N | 20 | 120 | 86 | 85 | 83 | 83 | 82 | 20 / –196 | 20 | 8–11 / 0.5 | 1.5–3 |
| | | | | | 100 | 86 | 80 | 77 | 76 | 76 | | | | |
| Mg–Th–Zr** | Th 3.0, Zr 0.7 (MTZ) | Sand cast and fully heat treated | U N U U | 20 / 200 250 | — | 74 | 68 | 65 | 63 | 62 | | | | |
| | | | | | — | 48 | 40 | 36 | 34 | 32 | | | | |
| | | | | | — | 74 | 68 | 60 | 59 | 58 | | | | |
| | | | | | 80 | 63 | 59 | 54 | 52 | 51 | | | | |

*continued overleaf*

**Table 3.13** (continued)

| Material | Nominal* composition % | Condition | ‡ State | temp. °C | Fatigue strength† at specified cycles | | | | | | Impact strength§ for single blow fracture | | |
|---|---|---|---|---|---|---|---|---|---|---|---|---|---|
| | | | | | $10^5$ MPa | $5\times10^5$ MPa | $10^6$ MPa | $5\times10^6$ MPa | $10^7$ MPa | $5\times10^7$ MPa | Test temp. °C | Unnotched J | Notched J |
| Mg–Th–Zn–Zr*** | Th 0.7, Zn 0.5, Zr 0.6 (ZTY) | Extruded | U | 20 | 100 | 86 | 83 | 79 | 76 | 74 | | | |
| | | | N | | 73 | 52 | 51 | 49 | 48 | 46 | | | |
| | | | U | 200 | – | 74 | 68 | 60 | 59 | 57 | | | |
| | | | U | 250 | 80 | 63 | 59 | 54 | 52 | 51 | | | 1.5–3 |
| | Th 3.0, Zn 2.2, Zr 0.7 (ZT1) | Sand cast and heat treated | N | 20 | 97 | 82 | 79 | 74 | 71 | 68 | 20 | 7–8 | |
| | | | U | | 76 | 59 | 56 | 51 | 49 | 48 | | | |
| | | | U | 200 | 71 | 60 | 59 | 54 | 52 | 51 | –196 | 0.8 | |
| | | | U | 250 | 66 | 51 | 46 | 43 | 42 | 39 | | | |
| | | | U | 325 | – | – | 43 | 37 | 34 | 29 | | | |
| Mg–RE(D)–As–Zr–Cu | RE(D) 2.2, Ag 1.5, Zr 0.6, Cu 0.07 | Sand cast and fully heat treated | U | 20 | 103 | 94 | 93 | 92 | 91 | 90 | | | |
| Mg–Y–RE(Δ)–YZr | Y 4.0, RE(Δ) 3.4, Zr 0.6 | Sand cast and fully heat treated | U | 20 | 114 | 101 | 98 | 94 | 93 | 91 | | | |
| | | | U | 150 | 107 | 97 | 94 | 87 | 85 | 83 | | | |
| | | | U | 250 | 107 | 81 | 74 | 65 | 64 | 62 | | | |
| | Y 5.9, RE(Δ) 3.0, Zr 0.6 | Sand cast and fully heat treated | U | 20 | 113 | 104 | 102 | 100 | 99 | 97 | | | |
| | | | U | 200 | 118 | 96 | 90 | 84 | 83 | 82 | | | |
| | | | U | 250 | 115 | 84 | 78 | 67 | 66 | 65 | | | |
| Mg–Zn–Cu–Mn | Zn 6.0, Cu 2.7, Mn 0.5 | Sand cast and fully heat treated | U | 20 | – | – | 100 | 94 | 92 | 90 | | | |
| | | | N | | – | – | 62 | 57 | 56 | 55 | | | |

*It is usual to add 0.2–0.4% Mn to alloys containing aluminium to improve corrosion resistance.

**Solution heat treated in an atmosphere of hydrogen.

† Wohler rotating beam tests at 2960 c.p.m.

‡ U = Unnotched.

N = Notched. Semi-circular notch of 0.12 cm (0.047 in) radius. Stress concentration factor 1.8.

§ Hounsfield balanced impact test, notched bar values are equivalent to Izod values.

RE(D) = Neodymium enriched mischmetal.

***Thorium-containing alloys are being replaced by alternative Mg alloys.

RE(Δ) = Neodymium plus Heavy Rare Earths.

**Table 3.14**  HEAT TREATMENT OF MAGNESIUM ALLOY CASTINGS

Heat treatment conditions for magnesium sand castings can be varied depending on the particular components and specific properties required. The following are examples of the conditions used for each alloy which will give properties meeting current national and international specifications.

| Material | | Nominal* composition % | | Condition | Time h | Temperature °C |
|---|---|---|---|---|---|---|
| Mg–Al–Zn | (AZ80) | Al | 8.0 | TB | 12–24 | 400–420 |
| | | Zn | 0.4 | | | |
| | (AZ91) | Al | 9.5 | TB | 16–24 | 400–420 |
| | | Zn | 0.4 | | | |
| | (AZ91) | | | TF | 16–24 | 400–420 |
| | | | | | | Air cool |
| | | | | | 8–16 | 180–210 |
| Mg–Zn–Zr | (Z5Z) | Zn | 4.5 | TE | 10–20 | 170–200 |
| | | Zr | 0.7 | | | |
| Mg–Zn–RE–Zr | (RZ5) | Zn | 4.0 | TE | 2–4 | 320–340 |
| | | RE | 1.2 | | | Air cool |
| | | Zr | 0.7 | | 10–20 | 170–200 |
| | (ZRE1) | RE | 2.7 | TE | 10–20 | 170–200 |
| | | Zn | 2.2 | | | |
| | | Zr | 0.7 | | | |
| Mg–Th–Zr† | (HK31) | Th | 3.0 | TF | 2–4 | 560–570 |
| | | Zr | 0.7 | | | Air cool |
| | | | | | 10–20 | 195–205 |
| Mg–Zn–Th–Zr† | (TZ6) | Zn | 5.5 | TE | 2–4 | 320–340 |
| | | Th | 1.8 | | | Air cool |
| | | Zr | 0.7 | | 10–20 | 170–200 |
| | (ZT1) | Th | 3.0 | TE | 10–20 | 310–320 |
| | | Zn | 2.2 | | | |
| | | Zr | 0.7 | | | |
| Mg–Ag–RE(D)Zr | (QE22) | Ag | 2.5 | TF | 4–12 | 520–530 |
| | | RE(D) | 2.0 | | Water/Oil Quench | |
| | | Zr | 0.6 | | 8–16 | 195–205 |
| Mg–RE(D)–Ag– Zr–Cu | (EQ21) | RE(D) | 2.2 | TF | 4–12 | 515–525 |
| | | Ag | 1.5 | | Water/Oil Quench | |
| | | Zr | 0.6 | | 12–16 | 195–205 |
| | | Cu | 0.07 | | | |
| Mg–Ag–RE(D)– Th–Zr | (QH21)† | Ag | 2.5 | TF | 4–12 | 520–530 |
| | | RE(D) | 1.0 | | Water/Oil Quench | |
| | | Th | 1.0 | | 12–20 | 195–205 |
| | | Zr | 0.6 | | | |

**Table 3.14**    (*continued*)

| Material | | Nominal* composition % | | Condition | Time h | Temperature °C |
|---|---|---|---|---|---|---|
| Mg–Y–RE(Δ)–Zr | (WE43) | Y | 4.0 | TF | 4–12 | 520–530 |
| | | RE(Δ) | 3.4 | | Water/Oil Quench | |
| | | Zr | 0.6 | | 12–20 | 245–255 |
| | (WE54) | Y | 5.1 | TF | 4.12 | 520–530 |
| | | RE(Δ) | 3.0 | | Water/Oil Quench/Air Cool | |
| | | Zr | 0.6 | | 12–20 | 245–255 |
| Mg–Zn–Cu–Mn | (ZC63) | Zn | 6.0 | TF | 4–12 | 435–445 |
| | | Cu | 2.7 | | Water Quench | |
| | | Mn | 0.5 | | 16–24 | 180–200 |

Note:- Above 350 °C, furnace atmospheres must be inhibited to prevent oxidation of magnesium alloys. This can be achieved either by:
(i) adding 1/2–1%$SO_1$ gas to the furnace atmosphere; or
(ii) carrying out the heat treatment in an atmosphere of 100% dry $CO_2$.
*It is usual to add 0.2–0.4% Mn to alloys containing aluminium to improve corrosion resistance.
　RE = Cerium mischmetal containing approximately 50% cerium.　　TB = Solution heat treated.
　RE(D) = Neodymium-enriched mischmetal.　　　　　　　　　　　TE = Precipitation heat treated.
　RE(Δ) = Neodymium plus Heavy Rare Earth metals.　　　　　　　TF = Fully heat treated.
†Thorium-containing alloys are being replaced by alternative Mg alloys.

*Mechanical properties at subnormal temperatures*

At temperatures down to −200 °C tensile properties have approximately linear temperature coefficients: proof stress and UTS increase by 0.1–0.2% of the RT value per °C fall in temperature, and elongation falls at the same rate: modulus of elasticity rises approximately 19 MPa (2800 lbf in$^{-2}$) per °C over the range 0° to −100 °C. No brittle-ductile transitions have been found.

　　Tests at −70 °C have suggested that the magnesium-zinc-zirconium alloys show the best retention of ductility and notched impact resistance at this temperature.

## 3.3 Mechanical properties of titanium and titanium alloys

**Table 3.15** PURE TITANIUM, TYPICAL MECHANICAL PROPERTIES AT ROOM TEMPERATURE

| Designation* | Grade | Condition | 0.2% proof stress MPa | Tensile strength MPa | Elongation on 50 mm | Elongation on 5D | Red. in area % | Specification bend radius 180° bend <1.83 mm | <3.25 mm | Mod. of elasticity GPa | Mod. of rigidity GPa |
|---|---|---|---|---|---|---|---|---|---|---|---|
| Iodide | Pure, 60 HV | | 103 | 241 | 55 | | 80 | | | | |
| IMI 115 | Commercially pure | Annealed sheet | 255 | 370 | 33 | | | | | | |
| | | Annealed rod | 220 | 370 | | 40 | 70 | 1t | 2t | | |
| | | Annealed wire | | 390 | 38 | | | | | | |
| IMI 125 | Commercially pure | †Annealed sheet | 340 | 460 | 30 | | | $1\frac{1}{2}$t | 2t | | |
| | | Annealed rod | 305 | 460 | | 28 | 57 | | | | |
| | | Annealed tube | 325 | 480 | 35 | | | | | | |
| IMI 130 | Commercially pure | Annealed sheet | 420 | 540 | 25 | | | $2$t$^+$ | $2\frac{1}{2}$t | | |
| | | Annealed rod | 360 | 540 | | 24 | 48 | | | 105 | 38 |
| | | Annealed wire | | 550 | 24 | | | | | | |
| | | Hard-drawn wire | | 700 | 11.5 | | | | | | |
| IMI 155 | Commercially pure | Annealed sheet | 540 | 640 | 24 | | | $2\frac{1}{2}$t | 3t | | |
| IMI 160 | Commercially pure | Annealed rod | 500 | 670 | | 23 | 46 | | | | |
| | | Annealed wire | | 690 | 24 | | | | | | |

*IMI Nomenclature.   †Up to 16.3 mm.

**Table 3.16** TITANIUM ALLOYS TYPICAL MECHANICAL PROPERTIES AT ROOM TEMPERATURE

| Designation* | Nominal composition % | | Condition | 0.2% proof stress MPa | Tensile strength MPa | Elongation % | | Red. in area % | Specification bend radius 180° | Mod. of elasticity GPa | Mod. of rigidity GPa |
|---|---|---|---|---|---|---|---|---|---|---|---|
| | | | | | | on 50 mm | on 5D | | | | |
| IMI 230 | Cu | 6.0 | Annealed sheet | 520 | 620 | 24 | | | 2t(0.5–3 mm) | 125 | |
| | | | Aged sheet | 670 | 770 | 20 | | | 2t(typical) | | |
| | | | Annealed rod | 500 | 630 | | 27 | 45 | | 125 | |
| | | | Aged rod | 580 | 740 | | 22 | 41 | | | |
| IMI 260 | Pd | 0.2 | Similiar to commercially Pure Titanium 115 | | | | | | | | |
| IMI 261 | Pd | 0.2 | Similiar to commercially Pure Titanium 125 | | | | | | | | |
| IMI 315 | Al | 2.0 | Annealed rod | 590 | 720 | | 21 | 50 | | 120 | |
| | Mn | 2.0 | | | | | | | | | |
| IMI 317 | Al | 5.0 | Annealed sheet | 820 | 860 | 16 | | | 4t(<2 mm) 4½t(≤ 3 mm) | 120 | |
| | Sn | 2.5 | Annealed rod | 930 | 1 000 | | 15 | 37 | | | |
| IMI 318 | Al | 6.0 | Annealed sheet | 1 110 | 1 160 | 10 | | 40 | 5t(≤ 3.25 mm) | 106 | 46 |
| | V | 4.0 | Annealed rod | 990 | 1 050 | | 15 | 40 | | | |
| | | | Aged rod (fastener stock) | 1 050 | 1 140 | | 15 | | | | |
| | | | Hard-drawn wire | | 1 410 | 4 | | | | | |
| IMI 550 | Al | 4.0 | F.h.t. rod | 1 070 | 1 200 | | 14 | 42 | | 116 | |
| | Mo | 4.0 | | | | | | | | | |
| | Sn | 2.0 | | | | | | | | | |
| | Si | 0.5 | | | | | | | | | |
| IMI 551 | Al | 4.0 | F.h.t.rod | 1 140 | 1 300 | | 12 | 40 | | 113 | 43 |
| | Mo | 4.0 | | | | | | | | | |
| | Sn | 4.0 | | | | | | | | | |
| | Si | 0.5 | | | | | | | | | |

*continued overleaf*

| | Composition | Condition | | | | | | |
|---|---|---|---|---|---|---|---|---|
| IMI 679 | Sn 11.0<br>Al 2.25<br>Mo 1.0<br>Si 0.2 | Quenched and aged rod<br>Air-cooled and aged rod | 1080<br>1000 | 1230<br>1120 | 11<br>13 | 40<br>45 | 108 | 46 |
| IMI 680 | Sn 11.0<br>Mo 4.0<br>Al 2.25<br>Si 0.2 | Quenched and aged rod<br>Furnace-cooled and aged rod | 1200<br>1080 | 1350<br>1160 | 12<br>14 | 37<br>47 | 115 | |
| IMI 685 | Al 6.0<br>Zr 5.0<br>Mo 0.5<br>Si 0.25 | F.h.t. rod | 920 | 1020 | 11 | 22 | 124 | 47 |
| IMI 829 | Al 5.5<br>Sn 3.5<br>Zr 3.0<br>Nb 1.0<br>Mo 0.3<br>Si 0.3 | F.h.t. rod | 848 | 965 | 12 | 22 | 120 | |
| IMI 834 | Al 5.8<br>Sn 4.0<br>Zr 3.5<br>Nb 0.7<br>Mo 0.5<br>Si 0.35<br>C 0.06 | F.h.t. rod | 931 | 1067 | 13 | 22 | 120 | |

*IMI Nomenclature.

**Table 3.17**  COMMERCIALLY PURE TITANIUM SHEET, TYPICAL VARIATION OF PROPERTIES WITH
TEMPERATURE

| Designation* | Temperature °C | 0.2% proof stress MPa | Tensile strength MPa | Elongation on 50 mm % | Mod. of elasticity GPa | Transformation temperature °C |
|---|---|---|---|---|---|---|
| IMI 115 | −196 | 442 | 641 | 34 | | $\alpha/\alpha + \beta$ |
| | −100 | 306 | 444 | 34 | | 865 |
| | 20 | 207 | 337 | 40 | | |
| | 100 | 168 | 296 | 43 | | |
| | 200 | 99 | 218 | 38 | | |
| | 300 | 53 | 167 | 47 | | |
| | 400 | 42 | 131 | 52 | | |
| | 450 | 36 | 120 | 49 | | |
| IMI 125 | 20 | 334 | 479 | 31 | | |
| | 100 | 250 | 397 | 32 | | |
| | 200 | 184 | 300 | 40 | | |
| | 300 | 142 | 232 | 45 | | |
| | 400 | 127 | 190 | 38 | | |
| | 450 | 119 | 175 | 35 | | |
| IMI 130 | −196 | 730 | 855 | 28 | | |
| | −100 | 590 | 737 | 28 | | |
| | 20 | 394 | 547 | 28 | 108 | |
| | 100 | 315 | 462 | 29 | 99 | $\alpha + \beta/\beta$ |
| | 200 | 205 | 331 | 37 | 91 | 915 |
| | 300 | 139 | 247 | 40 | 83 | |
| | 400 | 102 | 199 | 34 | 65 | |
| | 450 | 93 | 182 | 28 | | |
| | 500 | | | | 46 | |
| IMI 155 | 20 | 460 | 625 | 25 | | |
| | 100 | 372 | 537 | 26 | | |
| | 200 | 219 | 386 | 32 | | |
| | 300 | 151 | 281 | 36 | | |
| | 400 | 110 | 221 | 33 | | |
| | 450 | 96 | 202 | 26 | | |

*IMI nomenclature.

**Table 3.18**  TITANIUM ALLOYS, TYPICAL VARIATION OF PROPERTIES WITH TEMPERATURE

| Designation | Nominal composition % | | Condition | Temperature °C | 0.2% Proof stress MPa | Tensile strength MPa | Elongation % | | Red in area % | Mod. of elasticity GPa | Transformation temperature °C |
|---|---|---|---|---|---|---|---|---|---|---|---|
| | | | | | | | on 50 mm | on 5D | | | |
| IMI 230 | Cu | 2.5 | S.h.t (trans.) | 20 | 500 | 605 | 24 | | | | $\alpha/\alpha + \beta$ |
| | | | | 100 | 410 | 540 | 29 | | | | 790 |
| | | | | 200 | 310 | 450 | 33 | | | | |
| | | | | 300 | 270 | 410 | 31 | | | | |
| | | | | 400 | 250 | 380 | 30 | | | | |
| | | | | 500 | 220 | 380 | 33 | | | | |
| | | | Aged sheet (trans.) | 20 | 622 | 761 | 24 | | | | $\alpha + \beta/\beta$ |
| | | | | 100 | 553 | 704 | 23 | | | | $895 \pm 10$ |
| | | | | 200 | 471 | 635 | 26 | | | | |
| | | | | 300 | 457 | 607 | 23 | | | | |
| | | | | 400 | 429 | 573 | 19 | | | | |
| | | | | 500 | 357 | 468 | 21 | | | | |
| | | | Aged | 20 | 638 | 795 | | 22 | 40 | 107 | |
| | | | | 100 | 601 | 761 | | 21 | 39 | 100 | |
| | | | | 200 | 507 | 687 | | 23 | 45 | 92 | |
| | | | | 300 | 496 | 658 | | 20 | 50 | 85 | |
| | | | | 400 | 415 | 592 | | 21 | 53 | 78 | |
| | | | | 500 | 361 | 491 | | 27 | 57 | 71 | |
| IMI 260 | Pd | 0.2 | Similar to IMI 115 | | | | | | | | |
| IMI 262 | Pd | 0.2 | Similar to IMI 125 | | | | | | | | |
| IMI 315 | Al | 2.0 | Annealed rod | 20 | 618 | 757 | | 18 | 41 | 110 | $\alpha + \beta/\beta$ |
| | Mn | 2.0 | | 100 | 510 | 649 | | 21 | 46 | 107 | $915 \pm 20$ |
| | | | | 200 | 386 | 525 | | 22 | 48 | 97 | |
| | | | | 300 | 293 | 432 | | 19 | 50 | 86 | |
| | | | | 400 | 278 | 417 | | 18 | 56 | 76 | |
| | | | | 500 | 201 | 340 | | 22 | 72 | 62 | |
| IMI 317 | Al | 5.0 | Annealed rod | 20 | 822 | 919 | | 18 | 39 | 112 | $\alpha/\alpha + \beta$ |
| | Sn | 2.5 | | 100 | 692 | 798 | | 19 | 40 | 109 | 950 |
| | | | | 200 | 494 | 638 | | 18 | 44 | 105 | $\alpha + \beta/\beta$ |
| | | | | 300 | 415 | 576 | | 19 | 42 | 89 | $1025 \pm 20$ |
| | | | | 400 | 374 | 522 | | 18 | 41 | 84 | |
| | | | | 500 | 346 | 485 | | 21 | 57 | 81 | |

**Table 3.18** *(continued)*

| Designation | Nominal composition % | | Condition | Temperature °C | 0.2% Proof stress MPa | Tensile strength MPa | Elongation % | | Red in area % | Mod. of elasticity GPa | Transformation temperature °C |
|---|---|---|---|---|---|---|---|---|---|---|---|
| | | | | | | | on 50 mm | on 5D | | | |
| IMI 318 | Al | 6.0 | Annealed rod | −196 | 1560 | 1675 | | 6 | 29 | | $\alpha + \beta/\beta$ |
| | V | 4.0 | | −100 | 1165 | 1265 | | 12 | 33 | | $1000 \pm 15$ |
| | | | | 20 | 970 | 1040 | | 15 | 38 | 106 | |
| | | | | 100 | 825 | 920 | | 17 | 43 | 102 | |
| | | | | 200 | 710 | 815 | | 18 | 49 | 96 | |
| | | | | 300 | 645 | 750 | | 18 | 56 | 90 | |
| | | | | 400 | 580 | 700 | | 18 | 63 | 85 | |
| | | | | 500 | 450 | 605 | | 26 | 72 | 79 | |
| | | | | 600 | 125 | 265 | | 58 | 85 | | |
| | | | | 700 | 40 | 135 | | 127 | 94 | | |
| | | | Heat-treated rod (fastener stock) | 20 | 1035 | 1145 | | 14 | | | |
| | | | | 100 | 925 | 1035 | | 15 | | | |
| | | | | 200 | 805 | 925 | | 16 | | | |
| | | | | 300 | 710 | 850 | | 16 | | | |
| | | | | 400 | 635 | 805 | | 18 | | | |
| | | | | 500 | 540 | 695 | | 25 | | | |
| IMI 550 | Al | 4.0 | F.h.t. rod | 20 | 1081 | 1220 | | 15 | 49 | 116 | $\alpha + \beta/\beta$ |
| | Mo | 4.0 | | 100 | 965 | 1130 | | 15 | 49 | 112 | $980 \pm 10$ |
| | Sn | 2.0 | | 200 | 805 | 960 | | 16 | 60 | 106 | |
| | Si | 0.5 | | 300 | 700 | 900 | | 16 | 55 | 101 | |
| | | | | 400 | 655 | 835 | | 17 | 60 | 95 | |
| | | | | 500 | 585 | 780 | | 19 | 68 | 90 | |
| | | | | 600 | 310 | 585 | | 26 | 83 | 85 | |
| IMI 551 | Al | 4.0 | F.h.t rod | 20 | 1250 | 1390 | | 10 | 27 | 113 | $\alpha + \beta/\beta$ |
| | Mo | 4.0 | | 100 | 1125 | 1300 | | 11 | 29 | 108 | $1050 \pm 15$ |
| | Sn | 4.0 | | 200 | 925 | 1145 | | 14 | 38 | 103 | |
| | Si | 0.5 | | 300 | 815 | 1045 | | 15 | 38 | 98 | |
| | | | | 400 | 745 | 970 | | 14 | 41 | 93 | |
| | | | | 500 | 670 | 920 | | 18 | 55 | 88 | |
| | | | | 600 | 460 | 755 | | 27 | 65 | 81 | |
| IMI 679 | Sn | 11.0 | Quenched and aged rod | 20 | 1050 | 1230 | | 10 | 37 | | $\alpha + \beta/\beta$ |
| | Zr | 5.0 | | 100 | 940 | 1145 | | 11 | 43 | | $950 \pm 10$ |
| | Al | 2.25 | | 200 | 820 | 1020 | | 12 | 45 | | |
| | Mo | 1.0 | | 300 | 740 | 990 | | 11 | 46 | | |
| | Si | 0.2 | | 400 | 710 | 940 | | 11 | 46 | | |
| | | | | 450 | 680 | 910 | | 11 | 46 | | |

| Alloy | Composition (%) | | Condition | Temp | | | | | | Note |
|---|---|---|---|---|---|---|---|---|---|---|
| IMI 680 | Sn<br>Mo<br>Al<br>si | 11.0<br>4.0<br>2.25<br>0.2 | Air-cooled and aged rod | 20<br>100<br>200<br>300<br>400<br>500 | 1020<br>895<br>770<br>695<br>665<br>600 | 1095<br>995<br>900<br>865<br>850<br>795 | 14<br>16<br>16<br>14<br>14<br>15 | 41<br>47<br>49<br>49<br>48<br>48 | 108<br>103<br>99<br>94<br>90<br>85 | $\alpha + \beta/\beta$<br>945 ± 15 |
| IMI 685 | Al<br>Zr<br>Mn<br>Si | 6.0<br>5.0<br>0.5<br>0.25 | Quenched and aged rod | 20<br>100<br>200<br>300<br>400<br>450 | 1180<br>1020<br>905<br>835<br>805<br>725 | 1330<br>1190<br>1105<br>1075<br>1020<br>975 | 12<br>14<br>15<br>15<br>14<br>13 | 43<br>49<br>53<br>56<br>57<br>54 | 106<br>100<br>96<br>94<br>90<br>88 | $\alpha + \beta/\beta$<br>1020 ± 10 |
| | | | Furnace-cooled and aged rod | −196<br>−100<br>20 | 1630<br>1280<br>1030 | 1730<br>1380<br>1130 | 8½<br>10<br>15 | 36<br>43<br>49 | | |
| | | | F.h.t. rod | −196<br>−100<br>20<br>100<br>200<br>300<br>400<br>500 | 1480<br>1140<br>890<br>800<br>720<br>650<br>595<br>535 | 1560<br>1270<br>1030<br>935<br>850<br>800<br>750<br>695 | 6<br>10<br>12<br>13<br>15<br>16<br>18<br>19 | 13<br>18<br>22<br>22<br>24<br>27<br>31<br>37 | 124<br>120<br>114<br>108<br>102<br>95 | |
| IMI 829 | Al<br>Sm<br>Zr<br>Nb<br>Mo<br>Si | 5.5<br>3.5<br>3.0<br>1.0<br>0.3<br>0.3 | F.h.t. rod | 20<br>200<br>500<br>540<br>600 | 895<br>622<br>501<br>487<br>457 | 1028<br>792<br>665<br>653<br>634 | 10½<br>14½<br>15<br>16<br>14 | 22<br>28<br>36<br>42<br>38 | 119<br>110<br>93<br>91<br>88 | $\alpha + \beta\beta$<br>1015 ± 15 |
| IMI 834 | Al<br>Sn<br>Zr<br>Nb<br>Mo<br>Si<br>C | 5.8<br>4.0<br>3.5<br>0.7<br>0.5<br>0.35<br>0.06 | F.h.t. rod | 20<br>100<br>200<br>300<br>400<br>500<br>600 | 931<br>840<br>746<br>700<br>662<br>609<br>505 | 1067<br>962<br>885<br>832<br>790<br>764<br>656 | 13<br>13<br>14<br>14<br>14<br>15<br>16 | 22<br>23<br>27<br>32<br>36<br>42<br>50 | 120<br>116<br>112<br>106<br>102<br>96<br>92 | 1045 |

**Table 3.19**  COMMERCIALLY PURE TITANIUM – TYPICAL CREEP PROPERTIES

| IMI designation | Temperature °C | Stress MPa *to produce* 0.1% *plastic strain in* | | |
|---|---|---|---|---|
| | | 1000 h | 10 000 h | 100 000 h |
| IMI 130 | 20 | 288 | 270 | 207 |
| | 50 | 243 | 221 | 165 |
| | 100 | 179 | 165 | 119 |
| | 150 | 140 | 133 | 96 |
| | 200 | 113 | 116 | 77 |
| | 250 | 96 | 101 | 66 |
| | 300 | 87 | 83 | 55 |
| IMI 155 | 20 | 309 | 278 | 260 |
| | 50 | 252 | 232 | 213 |
| | 100 | 188 | 170 | 157 |
| | 150 | 145 | 131 | 122 |
| | 200 | 116 | 108 | 104 |
| | 250 | 102 | 97 | 94 |
| | 300 | 93 | 90 | 86 |

**Table 3.20**  TITANIUM ALLOYS – TYPICAL CREEP PROPERTIES

| IMI designation | Nominal composition % | | Condition | Temperature °C | Stress MPa *to produce* 0.1% *total plastic strain in* | | | |
|---|---|---|---|---|---|---|---|---|
| | | | | | 100 h | 300 h | 500 h | 1000 h |
| IMI 230 | Cu | 2.5 | Aged sheet | 200 | 435 | – | – | – |
| | | | | 300 | 375 | – | – | – |
| | | | | 400 | 220 | – | – | – |
| | | | | 450 | 109 | – | – | – |
| | | | Annealed sheet | 20 | 360 | – | – | – |
| | | | | 100 | 279 | – | – | – |
| | | | | 200 | 235 | – | – | – |
| | | | | 300 | 202 | – | – | – |
| | | | | 400 | 125 | – | – | – |
| IMI 317 | Al | 5.0 | Annealed rod | 20 | 633 | 608 | – | 593 |
| | Sn | 2.5 | | 100 | 474 | 463 | – | 458 |
| | | | | 200 | 370 | – | – | 370 |
| | | | | 300 | 359 | – | – | 359 |
| | | | | 400 | 337 | – | – | 337 |
| | | | | 500 | 162 | 119 | – | 88 |
| IMI 318 | Al | 60 | Annealed rod | 20 | 832 | 818 | – | 788 |
| | V | 4.0 | | 100 | 704 | 680 | – | 676 |
| | | | | 200 | 638 | 636 | – | 635 |
| | | | | 300 | 576 | 568 | – | – |
| | | | | 400 | 287 | 144 | – | 102 |
| | | | | 500 | 32 | 18 | – | – |
| IMI 550 | Al | 4.0 | Fully heat- | 300 | 724 | 718 | – | 710 |
| | Mo | 4.0 | treated bar | 400 | 551 | 516 | – | 471 |
| | Sn | 2.0 | | 450 | 254 | 174 | – | 101 |
| | Si | 0.5 | | 500 | 82 | 51 | – | 31 |
| IMI 551 | Al | 4.0 | Fully heat- | 400 | 621 | 575 | 540 | 501 |
| | Mo | 4.0 | treated rod | 450 | 307 | 217 | – | – |
| | Sn | 4.0 | | | | | | |
| IMI 679 | Sn | 11.0 | Air-cooled and | 20 | 896 | 880 | – | 880 |
| | Zr | 5.0 | aged rod | 150 | 703 | 695 | – | 672 |
| | Al | 2.25 | | 300 | 664 | 664 | – | 649 |
| | Mo | 1.0 | | 400 | 579 | 571 | – | 526 |
| | Si | 0.2 | | 450 | 448 | 386 | – | 247 |
| | | | | 500 | 131 | 93 | – | 62 |

**Table 3.20** (*continued*)

| IMI designation | Nominal composition % | | Condition | Temperature °C | Stress MPa *to produce* 0.1% *total plastic strain in* | | | |
|---|---|---|---|---|---|---|---|---|
| | | | | | 100 h | 300 h | 500 h | 1000 h |
| IMI 680 | Sn | 11.0 | Quenched and aged rod | 20 | 1127 | 1112 | – | – |
| | Mo | 4.0 | | 150 | 945 | 942 | – | – |
| | Al | 2.25 | | 200 | 862 | 856 | – | – |
| | Si | 0.2 | | 300 | 804 | 788 | – | – |
| | | | | 400 | 555 | 540 | – | – |
| | | | | 450 | 298 | 209 | – | – |
| | | | | 500 | 88 | 51 | – | – |
| | | | Furnace-cooled and aged rod | 300 | 570 | – | – | – |
| | | | | 350 | 540 | – | – | – |
| | | | | 400 | 490 | – | – | – |
| IMI 685 | Al | 6.0 | Heat-treated forgings | 200 | 599 | – | 592 | 589 |
| | Zr | 5.0 | | 300 | 551 | – | 541 | 535 |
| | Mo | 0.5 | | 400 | 497 | – | 480 | 462 |
| | Si | 0.25 | | 450 | 461 | – | 431 | 426 |
| | | | | 500 | 408 | – | 340 | – |
| IMI 829 | Al | 5.5 | Fully heat treated rod | 450 | 478 | – | – | – |
| | Sn | 3.5 | | 500 | 420 | – | – | – |
| | Zr | 3.0 | | 550 | 300 | – | – | – |
| | Nb | 1.0 | | 600 | 130 | – | – | – |
| | Mo | 0.3 | | | | | | |
| | Si | 0.3 | | | | | | |
| IMI 834 | Al | 5.8 | Heat-treated forgings | 500 | 461 | – | – | – |
| | Sn | 4.0 | | 550 | 339 | – | – | – |
| | Zr | 3.5 | | 600 | 205 | – | – | – |
| | Nb | 0.7 | | | | | | |
| | Mo | 0.5 | | | | | | |
| | Si | 0.35 | | | | | | |
| | C | 0.06 | | | | | | |

**Table 3.21**  TITANIUM AND TITANIUM ALLOYS – TYPICAL FATIGUE PROPERTIES

| IMI designation | Nominal composition % | Condition | Temperature °C | Tensile strength MPa | Details of test | Endurance limit for $10^7$ cycles (stated) MPa |
|---|---|---|---|---|---|---|
| IMI 115 | Commercial purity | Annealed rod | Room | 354 | Rotating bend Smooth $K_t = 1$ | ± 193 |
| | | | | 354 | Notched $K_t = 3$ | ± 123 |
| IMI 125 | Commercial purity | Annealed rod | Room | 417 | Rotating bend Smooth $K_t = 1$ | ± 232 |
| | | | | 417 | Notched $K_t = 3$ | ± 154 |
| IMI 130 | Commercial purity | Annealed rod | Room | 550 | Rotating bend Smooth $K_t = 1$ | ± 270 |
| | | | | 550 | Notched $K_t = 2$ | ± 170 |
| | | | | 550 | Notched $K_t = 3.3$ | ± 170 |
| | | | | 550 | Direct stress (Zero mean) Smooth $K_t = 1$ | ± 263 |
| | | | | 550 | Notched $K_t = 1,5$ | ± 247 |
| | | | | 550 | Notched $K_t = 2$ | ± 170 |
| | | | | 550 | Notched $K_t = 3.3$ | ± 116 |
| | | | | 589 | Smooth $K_t = 1$ | ± 278 |
| | | | | 589 | Notched $K_t = 2$ | ± 147 |
| | | | | 589 | Notched $K_t = 3$ | ± 123 |
| | | | | 589 | Notched $K_t = 4$ | ± 116 |

*continued overleaf*

**Table 3.21**   (*continued*)

| IMI designation | Nominal composition % | | Condition | Tempera- ture °C | Tensile strength MPa | Details of test | Endurance limit for $10^7$ cycles stated) MPa |
|---|---|---|---|---|---|---|---|
| IMI 160 | Commercial purity | | Annealed rod | Room | 674 | Direct stress (Zero mean) Smooth $K_t$ | ± 376 |
| IMI 230 | Cu | 2.5 | Annealed sheet | Room | 564 | Reversed bend | ± 390 |
| | | | Aged sheet | room | 772 | Reversed bend | ± 490 |
| | | | Aged sheet | room | 761 | Direct stress (Zero minimum) Smooth $K_t = 1$ | 0→560 |
| | | | Annealed rod | room 400 | 598 | Rotating bend Smooth $K_t = 1$ Smooth $K_t = 1$ | ± 370 ± 150 |
| | | | Annealed | Room | 638 | Direct stress (Zero mean) Smooth $K_t = 1$ | ± 280 |
| | | | Aged rod | Room 400 | 700 – | Rotating bend Smooth $K_t = 1$ Smooth $K_t = 1$ | ± 450 ± 290 |
| | | | Aged rod | Room | 792 | Direct stress (Zero mean) Smooth $K_t = 1$ Notched $K_t = 3.3$ | ± 470 ± 200 |
| IMI 260 | Pd | 0.2 | Similar to IMI 115 | | | | |
| IMI 262 | Pd | 0.2 | Similar to IMI 125 | | | Rotating bend | Limits for this alloy $10^8$ cycles |
| IMI 317 | Al Sn | 5.0 2.5 | Annealed rod | Room | – | Smooth $K_t = 1.0$ Notched $K_t = 2.0$ Notched $K_t = 3.3$ | ± 371 ± 263 ± 239 |
| | | | | | | Direct stress (Zero mean) Smooth $K_t = 1.0$ Notched $K_t = 1.5$ Notched $K_t = 2.0$ Notched $K_t = 3.3$ | ± 433 ± 278 ± 201 ± 154 |
| IMI 318 | Al V | 6.0 4.0 | Annealed rod | Room | 960 960 | Rotating bend Smooth $K_t = 1$ Notched $K_t = 2.7$ | ± 470 ± 230 |
| | | | | | 1015 1015 | Direct stress (Zero minimum) Smooth $K_t = 1$ Notched $K_t = 1$ | 0→750 0→325 |
| IMI 550 | Al Mo Sn Si | 4.0 4.0 2.0 0.5 | Fully heat- treated rod | Room | 1180 1180 | Direct stress (Zero minimum) Smooth $K_t = 1$ Notched $K_t = 3$ | 0→850 0→350 |
| | | | | | | Rotating bend Rotating bend Smooth $K_t = 1$ Notched $K_t = 2.4$ | ± 587 ± 394 |
| IMI 551 | Al Mo Sn Si | 4.0 4.0 4.0 0.5 | Fully heat- treated rod | Room | – – | Rotating bend Smooth $K_t = 1$ Notched $K_t = 3.2$ | ± 750 ± 430 |

**Table 3.21**  *(continued)*

| IMI designation | Nominal composition % | | Condition | Temperature °C | Tensile strength MPa | Details of test | Endurance limit for $10^7$ cycles (stated) MPa |
|---|---|---|---|---|---|---|---|
| IMI 679 | Sn | 11.0 | Air-cooled and aged rod | Room | – | Rotating bend Smooth $K_t = 1.0$ | $\pm 641$* |
| | Zr | 5.0 | | 200 | – | Smooth $K_t = 1.0$ | $\pm 510$* |
| | Al | 2.25 | | 400 | – | Smooth $K_t = 1.0$ | $\pm 510$* |
| | Mo | 1.0 | | 450 | – | Smooth $K_t = 1.0$ | $\pm 556$ |
| | Si | 0.2 | | 500 | – | Smooth $K_t = 1.0$ | $\pm 495$ |
| | | | | | | Rotating bend | (Limits for $2 \times 10^7$ cycles) |
| IMI 680 | Sn | 11.0 | Quenched and aged rod | Room | 1272 | Smooth $K_t = 1$ | $\pm 710$ |
| | Mo | 4.0 | | | 1272 | Notched $K_t = 2$ | $\pm 340$ |
| | Al | 2.25 | | | 1272 | Notched $K_t = 3.3$ | $\pm 293$ |
| | Si | 0.2 | | | | Direct stress (Zero mean | (Limits for $2 \times 10^7$ cycles) |
| | | | | Room | 1272 | Smooth $K_t = 1$ | $\pm 695$ |
| | | | | | | Notched $K_t = 2$ | $\pm 371$ |
| | | | | | | Notched $K_t = 3.3$ | $\pm 232$ |
| | | | | | | Rotating bend | (Limits for $10^8$ cycles) |
| | | | | Room | – | Smooth $K_t = 1$ | $\pm 648$ |
| | | | | 200 | – | Smooth $K_t = 1$ | $\pm 495$ |
| | | | | 400 | – | Smooth $K_t = 1$ | $\pm 479$ |
| | | | Furnace-cooled rod | Room | 1100 | Direct stress (Zero Smooth $K_t = 1$ | $\pm 680$ |
| IMI 685 | Al | 6.0 | Fully heat-treated rod | 20 | – | Direct stress (Zero mean) Smooth $K_t = 1$ | $\pm 440$ |
| | Zr | 5.0 | | 450 | – | Smooth $K_t = 1$ | $\pm 300$ |
| | Mo | 0.5 | | 520 | – | Smooth $K_t = 1$ | $\pm 260$ |
| | Si | 0.25 | | | | Direct stress (Zero minimum) | |
| | | | | 450 | – | Smooth $K_t = 1$ | $0 \rightarrow 475$ |
| | | | | 520 | – | Smooth $K_t = 1$ | $0 \rightarrow 425$ |
| | | | Fully heat-treated forging | Room | – | Direct stress (Zero minimum) Smooth $K_t = 1$ | $0 \rightarrow 640$ |
| | | | | Room | – | Notched $K_t = 3.5$ | $0 \rightarrow 220$ |
| | | | | 475 | – | Smooth $K_t = 1$ | $0 \rightarrow 460$ |
| | | | | 475 | – | Notched $K_t = 3.5$ | $0 \rightarrow 210$ |
| IMI 829 | Al | 5.5 | Fully heat-treated rod | Room | – | Direct stress (Zero minimum) Smooth $K_t = 1$ | $0 \rightarrow 550$ |
| | Sn | 3.5 | | | – | Notched $K_t = 3$ | $0 \rightarrow 260$ |
| | Zr | 3.0 | | | | | |
| | Nb | 1.0 | | | | | |
| | Mo | 0.3 | | | | | |
| | Si | 0.3 | | | | | |
| IMI 834 | Al | 5.8 | Fully heat-treated rod | Room | – | Direct stress (Zero minimum) Smooth $K_t = 1$ | $0 \rightarrow 577$ |
| | Sn | 4.0 | | | – | Notched $K_t = 2$ | $0 \rightarrow 363$ |
| | Zr | 3.5 | | | | | |
| | Nb | 0.7 | | | | | |
| | Mo | 0.5 | | | | | |
| | Si | 0.35 | | | | | |
| | C | 0.06 | | | | | |

*Limits for $10^8$ cycles.

**Table 3.22**    IZOD IMPACT PROPERTIES OF TITANIUM AND TITANIUM ALLOYS

| *IMI* *designation* | *Nominal composition* *%* | | *Condition* | *Izod value* Joules (ft lbf)* | | | | | | | |
|---|---|---|---|---|---|---|---|---|---|---|---|
| | | | | −196 °C | −78 °C | 20 °C | 100 °C | 200 °C | 300 °C | 400 °C | 500 °C |
| IMI 130† | Commercially pure | | Annealed rod | – | 62.4 (46) | 61.0 (45) | 62.4 (46) | 72 (53) | 82 (60½) | 84 (62) | 82 (60½) |
| IMI 317 | Sn | 5.0 | Annealed rod | 17.6 (13) | 20.3 (15) | 27.1 (20) | 35.2 (26) | 52.8 (39) | 63.7 (47) | 70.5 (52) | 71.8 (53) |
| | Al | 2.5 | | | | | | | | | |
| IMI 318 | Al | 6.0 | Annealed rod | 13.5 (10) | 14.9 (11) | 20.3 (15) | 25.7 (19) | 40.6 (30) | 65.0 (48) | 83.5 (63) | 92.0 (68) |
| | V | 4.0 | | | | | | | | | |
| IMI 550 | Al | 4.0 | Fully heat-treated rod | – | – | 19.0 (14) | – | – | – | – | – |
| | Mo | 4.0 | | | | | | | | | |
| | Sn | 2.0 | | | | | | | | | |
| | Si | 0.5 | | | | | | | | | |

| *IMI* *designation* | *Nominal composition* *%* | | *Condition* | *Charpy value* Joules (ft lbf) | | | | | | | |
|---|---|---|---|---|---|---|---|---|---|---|---|
| | | | | −196 °C | −78 °C | 20 °C | 100 °C | 200 °C | 300 °C | 400 °C | 500 °C |
| IMI 551 | Al | 4.0 | Fully heat-treated rod | 13.5 (10) | 19 (14) | 20.3 (15) | 21.7 (16) | 24.4 (18) | 26.5 (19½) | 28.5 (21) | 31.2 (23) |
| | Mo | 4.0 | | | | | | | | | |
| | Sn | 4.0 | | | | | | | | | |
| | Si | 0.5 | | | | | | | | | |
| IMI 679 | Sn | 11.0 | Air-cooled and aged | 10.8 (8) | 13.5 (10) | 14.9 (11) | 16.3 (12) | 19 (14½) | 25 (18½) | 30 (22) | 33.9 (25) |
| | Zr | 5.0 | | | | | | | | | |
| | Al | 2.25 | | | | | | | | | |
| | Mo | 1.0 | | | | | | | | | |
| | Si | 0.2 | | | | | | | | | |
| IMI 680 | Sn | 11.0 | Quenched and aged rod | 8.1 (6) | 8.8 (6½) | 10.8 (8) | 12.2 (9) | 14.9 (11) | 17.6 (13) | 20.3 (15) | 25.7 (19) |
| | Mo | 4.0 | | | | | | | | | |
| | Al | 2.25 | | | | | | | | | |
| | Si | 0.2 | | | | | | | | | |
| IMI 685 | Al | 6.0 | Fully heat-treated rod | 31.2 (23) | 39.3 (29) | 43.4 (32) | – | – | – | – | – |
| | Zr | 5.0 | | | | | | | | | |
| | Mo | 0.5 | | | | | | | | | |
| | Si | 0.25 | | | | | | | | | |

*BSS 131 (1) 0.45 in diameter straight notched test pieces. †Izod values of commercial purity titanium are appreciably affected by variation in hydrogen content within commercial limits (0.008% maximum) in Ti 130 rod.

# 4 Aluminium and magnesium casting alloys

## 4.1  Aluminium casting alloys

**Table 4.1**  ALUMINIUM-SILICON ALLOYS

| Specification BS 1490: 1988 Related British Specifications | LM6M(Ge) BS L33 | LM20M(Ge) | LM9M(SP) | LM9TE(SP) | LM9TE(SP) | LM13TE(SP) | LM13TF(SP) | LM13TF7(SP) |
|---|---|---|---|---|---|---|---|---|
| *Composition (%) (single figure indicates maximum)* | | | | | | | | |
| Copper | 0.1 | 0.4 | | 0.1 | | | 0.7–1.5 | |
| Magnesium | 0.1 | 0.2–0.6 | | 0.2 | | | 0.8–1.5 | |
| Silicon | 10.0–13.0 | 10.0–13.0 | | 10.0–13.0 | | | 10.0–12.0 | |
| Iron | 0.6 | 1.0 | | 0.6 | | | 1.0 | |
| Manganese | 0.5 | 0.5 | | 0.3–0.7 | | | 0.5 | |
| Nickel | 0.1 | 0.1 | | 0.1 | | | 1.5 | |
| Zinc | 0.1 | 0.2 | | 0.1 | | | 0.5 | |
| Lead | 0.1 | 0.1 | | 0.1 | | | 0.1 | |
| Tin | 0.05 | 0.1 | | 0.05 | | | 0.1 | |
| Titanium | 0.2 | 0.2 | | 0.2 | | | 0.2 | |
| Other | – | – | | – | | | – | |
| *Properties of material* | | | | | | | | |
| Suitability for: | | | | | | | | |
| Sand casting | E | E* | | G | | | G | |
| Chill casting (gravity die) | E | E | | E | | | G | |
| Die casting (press die) | G | G | | G* | | | F* | |
| Strength at elevated temperature | P | P | | G | | | E | |
| Corrosion resistance | E | G | | E | | | G | |
| Pressure tightness | E | E | | G | | | F | |
| Fluidity | E | E | | G | | | G | |
| Resistance to hot shortness | E | E | | E | | | E | |
| Machinability | F | F | | F | | | F | |
| Melting range, C | 565–575 | 565–575 | | 550–575 | | | 525–560 | |
| Casting temperature range, C | 710–740 | 680–740 | | 690–740 | | | 680–760 | |
| Specific gravity | 2.65 | 2.68 | | 2.68 | | | 2.70 | |

| | (1) | (2) | (3) | (4) | (5) | (6) | (7) | (8) |
|---|---|---|---|---|---|---|---|---|
| *Heat treatment* | | | | | | | | |
| Solution temperature, °C | – | – | – | – | 520–535 | – | 515–525 | 515–525 |
| Solution time, h | – | – | – | – | 2–8 | – | 8 (minimum) | 8 (minimum) |
| Quench | – | – | – | – | Cold water | – | Water, 70–80 °C | Water, 70–80 °C |
| Precipitation temperature, °C | – | – | – | 150–170 | 150–170 | 160–180 | 160–180 | For pistons: 200–250 |
| Precipitation time, h | – | – | – | 16 (minimum) | 16 (minimum) | 4–16 | 4–16 | 4–6** |
| Stabilization temperature, °C | – | – | – | – | – | – | – | – |
| Stabilization time, h | – | – | – | – | – | – | – | – |
| *Special properties* | Suitable for thin and intricate castings, readily welded | Pressure casting alloy | – | Suitable for low-pressure casting | High strength and hardness | Low coefficient of expansion | Good bearing properties | Piston alloy |
| *Mechanical properties – sand cast – SI units (Imperial units in brackets)* | | | | | | | | |
| Tensile stress min., MPa (tonf in$^{-2}$) | 160(10.4) | – | – | 170(11.0) | 240(15.5) | – | 170(11.0) | 140(9.1) |
| Elongation min.% | 5 | – | – | 1.5 | 0–1 | – | 0.5 | 1 |
| Expected 0.2% proof stress, MPa (tonf in$^{-2}$) | 60–70 (3.9–4.5) | – | – | 110–130 (7.1–8.4) | 220–250 (14.2–16.2) | – | 160–190 (10.4–12.3) HB 100–150 | 130(8.4) HB 65–85 |
| *Mechanical properties – chill cast – SI units (Imperial units in brackets)* | | | | | | | | |
| Tensile stress min., MPa (tonf in$^{-2}$) | 190(12.3) | 190(12.3) | 190(12.3) | 230(14.9) | 295(19.1) | 210(13.6) | 280(18.1) | 200(12.9) |
| Elongation, min. % | 7 | 5 | 3 | 2 | 0–1 | 1 | 1 | 1 |
| Expected 0.2% proof stress, MPa (tonf in$^{-2}$) | 70–80 (4.5–5.2) | 70–80 (4.5–5.2) | 75–85 (4.9–5.5) | 150–170 (9.7–11.0) | 270–280 (17.5–18.1) | – HB90–120 | 270–300 (17.5–19.4) HB 100–150 | 190(12.3) HB 65–85 |

*Not normally used in this form.

†If Ti alone is used for grain refinement then Ti ≯ 0.05%.

‡Fully heat-treated.

§Refine with phosphorus – subject to examination under microscope.

**Or for such time to give required BHN.

*Notes*

Association of Light Alloy Refiners and Smelters Grading:

E – Excellent, F – Fair, G – Good, P – Poor, U – Unsuitable,

(Ge – General purpose alloy; SP – special purpose alloy as per BS 1490:1988).

**Table 4.1**  (*continued*)

| Specification BS 1490: 1988 Related British Specifications | LM18M(SP) | LM25M(Ge) | LM25TE(Ge) | LM25TB7(Ge) | LM25TF(Ge) | LM29TE(SP) | LM29TF(SP) |
|---|---|---|---|---|---|---|---|
| *Composition % (Single figure indicates maximum)* | | | | | | | |
| Copper | 0.1 | | 0.20 | | | 0.8–1.3 | |
| Magnesium | 0.1 | | 0.20–0.60 | | | 0.8–1.3 | |
| Silicon | 4.5–6.0 | | 6.5–7.0 | | | 22–25 | |
| Iron | 0.6 | | 0.5 | | | 0.7 | |
| Manganese | 0.5 | | 0.3 | | | 0.6 | |
| Nickel | 0.1 | | 0.1 | | | 0.8–1.3 | |
| Zinc | 0.1 | | 0.1 | | | 0.2 | |
| Lead | 0.1 | | 0.1 | | | 0.1 | |
| Tin | 0.05 | | 0.05 | | | 0.1 | |
| Titanium | 0.2 | | $0.2^{\dagger}$ | | | 0.2 | |
| Other | – | | – | | | Cr 0.6; Co 0.5, p§ | |
| *Properties of material* | | | | | | | |
| Suitability for: | | | | | | | |
| Sand casting | G | | G | | | P | |
| Chill casting (gravity die) | G | | E | | | F | |
| Die casting (press die) | G* | | G* | | | U | |
| Strength at elevated temperature | P | | E | | | G | |
| Corrosion resistance | E | | E | | | G | |
| Pressure tightness | E | | G | | | F | |
| Fluidity | G | | G | | | F | |
| Resistance to hot shortness | E | | G | | | E | |
| Machinability | F | | F | | | P | |
| Melting range, C | 565–625 | | 550–615 | | | 520–770 | |
| Casting temperature range, C | 700–740 | | 680–740 | | | At least 830 | |
| Specific gravity | 2.69 | | 2.68 | | | 2.65 | |

| | | | | | | | |
|---|---|---|---|---|---|---|---|
| *Heat treatment* | | | | | | | |
| Solution temperature, °C | – | – | – | 525–545 | 525–545 | – | 495–505 |
| Solution time, h | – | – | – | 4–12 | 4–12 | – | 4 |
| Quench | – | – | – | Water, 70–80°C | Water, 70–80°C | – | Air blast |
| Precipitation temperature, °C | 155–175 | – | 155–175 | 155–175 | 155–175 | – | 185 |
| Precipitation time, h | 8–12 | – | 8–12 | 8–12 | 8–12 | – | 8 |
| Stabilization temperature, °C | – | – | – | – | 250 | – | To produce HB requirement |
| Stabilization time, h | – | – | – | – | 2–4 | – | – |
| *Special properties* | | Readily welded | | | General purpose high-strength casting alloy | | More suited to chill (grav, die) casting / Piston alloy |
| *Mechanical properties – sand cast – SI units (Imperial units in brackets)* | | | | | | | |
| Tensile stress min., MPa (tonf in$^{-2}$) | 120(7.8) | 130(8.4) | 150(9.7) | 160(10.4) | 230(14.9) | 120(7.8) | 120(7.8) |
| Elongation min. % | 3 | 2 | 1 | 2.5 | 0–2 | 0.3 | 0.3 |
| Expected 0.2% proof stress, MPa (tonf in$^{-2}$) | 55–60(3.6–3.9) | 80–100(5.2–6.5) | 120–150(7.8–9.7) | 80–110(5.2–6.5) | 200–250(12.9–16.2) | 120(7.8) | 120(7.8) |
| MPa (tonf in$^{-2}$) | | | | HB 100–140 | | HB 100–140 | HB 100–140 |
| *Mechanical properties – chill cast – SI units (Imperial units in brackets)* | | | | | | | |
| Tensile stress min., MPa (tonf in$^{-2}$) | 140(9.1) | 160(10.4) | 190(12.3) | 230(14.9) | 280(18.1) | 190(12.3) | 190(12.3) |
| Elongation min. % | 4 | 3 | 2 | 5 | 2 | 0.3 | 0.3 |
| Expected 0.2% proof stress, MPa (tonf in$^{-2}$) | 60–70(3.9–4.5) | 80–100(5.2–6.5) | 130-200(8.4–12.9) | 90–110(5.8–7.1) | 220–260(14.2–16.8) | 170(11.0) | 170–190(11.0–12.3) |
| MPa (tonf in$^{-2}$) | | | | HB 100–140 | | HB 100–140 | HB 100–140 |

*Not normally used in this form.

†If Ti alone is used for grain refinement then Ti ≯ 0.05%.

‡Fully heat-treated.

§Refine with phosphorus-subject to examination under microscope.

**Or for such time to give required BHN.

Note

E – Excellent. F – Fair. G – Good. P – Poor. U – Unsuitable.

(Ge – General purpose alloy; SP – Special purpose alloy as per BS: 1490: 1988).

**Table 4.2**   ALUMINIUM-SILICON-COPPER ALLOYS

| Specification BS 1490; 1988 Related British Specifications | LM2M(Ge) | LM4M(Ge) | LM4MTF (Ge) | LM16TB (SP) | LM16TF (SP) 3L78 | LM21M(SP) |
|---|---|---|---|---|---|---|
| *Composition %* (Single figures indicate maximum) | | | | | | |
| Copper | 0.7–2.5 | 2.0–4.0 | | 1.0–1.5 | | 3.0–5.0 |
| Magnesium | 0.30 | 0.15 | | 0.4–0.6 | | 0.1–0.3 |
| Silicon | 9.0–11.5 | 4.0–6.0 | | 4.5–5.5 | | 5.0–7.0 |
| Iron | 1.0 | 0.8 | | 0.6 | | 1.0 |
| Manganese | 0.5 | 0.2–0.6 | | 0.5 | | 0.2–0.6 |
| Nickel | 0.5 | 0.3 | | 0.25 | | 0.3 |
| Zinc | 2.0 | 0.5 | | 0.1 | | 2.0 |
| Lead | 0.3 | 0.1 | | 0.1 | | 0.2 |
| Tin | 0.2 | 0.1 | | 0.05 | | 0.1 |
| Titanium | 0.2 | 0.2 | | 0.2* | | 0.2 |
| *Properties of material* Suitability for: | | | | | | |
| Sand casting | G† | G | | G | | G |
| Chill casting (gravity die) | G† | G | | G | | G |
| Die casting (press die) | E | G | | F† | | G† |
| Strength at elevated temp. | G‡ | G | | G | | G |
| Corrosion resistance | G | G | | G | | G |
| Pressure tightness | G | G | | G | | G |
| Fluidity | G | G | | G | | G |
| Resistance to hot shortness | E | G | | G | | G |
| Machinability | F | G | | G | | G |
| Melting range, °C | 525–570 | 525–625 | | 550–620 | | 520–615 |
| Casting temperature range, °C | – | 700–760 | | 690–760 | | 680–760 |
| Specific gravity | 2.74 | 2.73 | | 2.70 | | 2.81 |
| *Heat treatment* | | | | | | |
| Solution temperature, °C | – | – | 505–520 | 520–530 | 520–530 | – |
| Solution time, h | – | – | 6–16 | 12 (min) | 12 (min) | – |
| Quench | – | – | Water at 70–80°C | Water at 70–80°C | Water at 70–80°C | – |
| Precipitation temperature, °C | – | – | 150–170 | – | 160–170 | – |
| Precipitation time, h | – | – | 6–18 | – | 8–10 | – |
| *Special properties* | Alloy for pressure die castings | General engineering alloy Can tolerate relatively high static loading in TF condition | | Pressure tight. High strength alloy in TF condition | | Equally suited to all casting processes |
| *Mechanical properties – sand cast* – SI units (Imperial units in brackets) | | | | | | |
| Tensile stress min. MPa (tonf in$^{-2}$) | – | 140(9.1) | 230(14.9) | 170(11.0) | 230(14.9) | 150(9.7) |
| Elongation min. % | – | 2 | – | 2 | – | 1 |
| Expected 0.2% proof stress, MPa (tonf in$^{-2}$) | – | 70–110 (4.5–7.1) | 200–250 (12.9–16.2) | 120–140 (7.8–9.1) | 220–280 (14.2–18.1) | 80–140 (5.2–9.1) |
| *Mechanical properties – chill cast* – SI units (Imperial units in brackets) | | | | | | |
| Tensile strength min. MPa (tonf in$^{-2}$) | 150(9.7) | 160(10.4) | 280(18.1) | 230(14.9) | 280(18.1) | 170(11.0) |
| Elongation min. % | 1 | 2 | 1 | 3 | – | 1 |
| Expected 0.2% proof stress, MPa (tonf in$^{-2}$) | 90–130 (5.8–8.4) | 80–110 (5.2–7.1) | 200–300 (12.9–19.4) | 140–150 (9.1–9.7) | 250–300 (16.2–19.4) | 80–140 (5.2–9.1) |

*0.05% min. if Ti alone used for grain refinement.

†Not normally used in this form.

‡The use of die castings is usually restricted to only moderately elevated temperatures.

**Table 4.2**  (*continued*)

| *Specification* BS 1490: 1988<br>*Related British Specifications* | LM22TB<br>(SP) | LM24M<br>(Ge) | LM26TE<br>(SP) | LM27M<br>(Ge) | LM30M<br>(SP) | LM30TS<br>(SP) |
|---|---|---|---|---|---|---|
| *Composition %* (Single figures indicate maximum) | | | | | | |
| Copper | 2.8–3.8 | 3.0–4.0 | 2.0–4.0 | 1.5–2.5 | 4.0–5.0 | |
| Magnesium | 0.05 | 0.1 | 0.5–1.5 | 0.3 | 0.4–0.7 | |
| Silicon | 4.0–6.0 | 7.5–9.5 | 8.5–10.5 | 6.0–8.0 | 16–18 | |
| Iron | 0.6 | 1.3 | 1.2 | 0.8 | 1.1 | |
| Manganese | 0.2–0.6 | 0.5 | 0.5 | 0.2–0.6 | 0.3 | |
| Nickel | 0.15 | 0.5 | 1.0 | 0.3 | 0.1 | |
| Zinc | 0.15 | 3.0 | 1.0 | 1.0 | 0.2 | |
| Lead | 0.1 | 0.3 | 0.2 | 0.2 | 0.1 | |
| Tin | 0.05 | 0.2 | 0.1 | 0.1 | 0.1 | |
| Titanium | 0.2 | 0.2 | 0.2 | 0.2 | 0.2 | |
| *Properties of material*<br>Suitability for: | | | | | | |
| Sand casting | G† | F† | G | G | U | |
| Chill casting (gravity die) | G | F† | G | E | F | |
| Die casting (press die) | G† | E | F† | G† | G | |
| Strength at elevated temp. | G | G‡ | E | G | G | |
| Corrosion resistance | G | G | G | G | G | |
| Pressure tightness | G | G | F | G | F | |
| Fluidity | G | G | G | G | G | |
| Resistance to hot shortness | G | G | F | G | F | |
| Machinability | G | F | F | G | P | |
| Melting range, °C | 525–625 | 520–580 | 520–580 | 525–605 | 505–650 | |
| Casting temperature range, °C | 700–740 | – | 670–740 | 680–740 | Well above 650°C | |
| Specific gravity | 2.77 | 2.79 | 2.76 | 2.75 | 2.73 | |
| *Heat treatment* | | | | | | |
| Solution temperature, °C | 515–530 | – | – | – | – | |
| Solution time, h | 6–9 | – | – | – | – | |
| Quench | Water at<br>70–80 °C | – | – | – | – | |
| Precipitation temperature, °C | – | – | 200–210 | – | – | *Stress relief*<br>175–225 |
| Precipitation time, h | – | – | 7–9 | – | – | 8(minimum) |
| *Special properties* | Chill casting<br>alloy (grav.<br>die) | Alloy for<br>pressure<br>die castings | Piston alloy,<br>retains<br>strength<br>and<br>hardness at<br>elevated<br>temps. | Excellent<br>castability | Alloy for<br>pressure die<br>casting<br>cylinder blocks | Alloy for pressure die<br>automobile engine |
| *Mechanical properties – sand cast* – SI units (Imperial units in brackets)<br>Tensile stress min., MPa<br>(tonf in⁻²) | – | – | – | 140(9.1) | – | – |
| Elongation min. % | – | – | – | 1 | – | – |
| Expected 0.2% proof stress,<br>MPa (tonf in⁻²) | – | – | – | 80–90<br>(5.2–5.8) | – | – |

*Mechanical properties – chill cast* – SI units (Imperial units in brackets)
HB = 90–120

| Tensile strength min., MPa<br>(tonf in⁻²) | 245(15.9) | 180(11.7) | 210(13.6) | 160(10.4) | 150(9.7) | 160(10.4) |
|---|---|---|---|---|---|---|
| Elongation min. % | 8 | 1.5 | 1 | 2 | 0.5 | 0.5 |
| Expected 0.2% proof stress,<br>MPa (tonf in⁻²) | 110–120<br>(7.1–7.8) | 100–120<br>(6.7–7.7) | 160–190<br>(10.4–12.3) | 90–110<br>(5.8–7.1) | 150–200<br>(9.7–12.9) | 160–200<br>(10.4–12.9) |

*Note:*
E – Excellent. F – Fair. G – Good. P – Poor. U – Unsuitable.
(Ge – General purpose alloy; Sp-Special purpose alloy as per BS 1490;1988).

**Table 4.3**    ALUMINIUM-COPPER ALLOYS

| Specification BS 1490; 1988 | LM12M(SP) | LM12TF(SP)* | [LM14-WP]† | [LM11-W] | [LM11-WP] | | |
|---|---|---|---|---|---|---|---|
| Aerospace | — | — | 4L35 | 2L91 | 2L92 | — | — |
| BSL series | — | — | — | — | — | 361B | 741A |
| DTD series | — | — | — | — | — | — | — |
| *Composition %* (Single figures indicate maximum) | | | | | | | |
| Copper | 9.0–11.0 | 9.0–11.0 | 3.5–4.5 | 4.0–5.0 | 4.0–5.0 | 4.0–5.0 | 3.5–4.5 |
| Magnesium | 0.2–0.4 | | 1.2–1.7 | 0.10 | 0.10 | 0.10 | 1.2–2.5 |
| Silicon | 2.5 | | 0.6‡ | 0.25 | 0.25 | 0.25 | 0.5 |
| Iron | 1.0 | | 0.6‡ | 0.25 | 0.25 | 0.25 | 0.5 |
| Manganese | 0.6 | | 0.6 | 0.10 | 0.10 | 0.10 | 0.1 |
| Nickel | 0.5 | | 1.8–2.3 | 0.10 | 0.10 | 0.10 | 0.1 |
| Zinc | 0.8 | | 0.1 | 0.10 | 0.10 | 0.05 | 0.1 |
| Lead | 0.1 | | 0.05 | 0.05 | 0.05 | 0.05 | 0.1 |
| Tin | 0.1 | | 0.05 | 0.05 | 0.05 | 0.05 | 0.05 |
| Titanium | 0.2 | | 0.25 | 0.25 | 0.25 | Ti + Nb 0.05–0.30 | — |
| Other | — | | — | — | — | — | Co 0.5–1.0  Nb 0.05–0.3 |
| *Properties of material* | | | | | | | |
| Suitability for: | | | | | | | |
| Sand casting | F | | F | F | F | F | F |
| Chill casting (gravity die) | G | | G | P | P | P | G |
| Die casting (press die) | U | | U | U | U | U | — |
| Strength at elevated temperature | G | | E | F | F | F | F |
| Corrosion resistance | P | | F | P | P | P | F |
| Pressure tightness | G | | E | F | F | F | F |
| Fluidity | F | | G | P | P | P | G |
| Resistance to hot shortness | G | | G | G | G | G | G |
| Machinability | E | | G | G | G | G | G |
| Melting range, C | 525–625 | | 530–640 | 545–640 | 545–640 | 540–650 | 530–640 |
| Casting temperature range, C | 700–760 | | 700–750 | 680–700 | 680–700 | 675–750 | 710–725 |
| Specific gravity | 2.94 | | 2.82 | 2.80 | 2.80 | 2.80 | 2.80 |

*Heat treatment*

| | | | | | | |
|---|---|---|---|---|---|---|
| Solution temp., °C | 515–520 | 500–520 | 525–545 | 525–545 | 525–545 | 495–505 |
| Solution time, h | 6 | 6 | 12–16 | 12–16 | 16 (minimum) | 10 |
| Quench | Water at 70–80°C | Boiling water | Water at 70–80°C | Water at 70–80°C | Water or oil | (minimum)†† Oil at 80–90°C |
| Precipitation temperature, °C | 175–180 | 95–103§ | 120–140 | 120–170 | 160–170 | 195–205 |
| Precipitation time, h | 2 (minimum) | 2** | 1–2 | 12–14 | 8–16 | 4–5 |
| *Special properties* | Piston alloy, now superseded by LM13 and LM26. Excellent machinability | Excellent props. at elevated temperatures. Grav. die alloy | Good shock resistance | | High strength alloy | |

*Mechanical properties – sand cast – SI (Imperial units in brackets)*

| | | | | | | |
|---|---|---|---|---|---|---|
| Tensile stress min., MPa (tonf in$^{-2}$) | 170 | 220(14.2) | 220(14.2) | 280(18.1) | 324(21.0) | 263(17.0) |
| Elongation % | – | – | 7 | 4 | – | – |
| Expected 0.2% proof stress, min., MPa (tonf in$^{-2}$) | 139–170 | 210–240 (13.6–15.5) | 165–200 (10.7–12.9) | 200–240 (12.9–15.5) | 310(20.1) | 250(16.2) |

*Mechanical properties – chill cast – SI units (Imperial units in brackets)*

| | | | | | | |
|---|---|---|---|---|---|---|
| Tensile stress min., MPa (tonf in$^{-2}$) | 278(18.0) | 280(18.1) | 265(17.1) | 310(20.1) | 402(26.0) | 340(22.0) |
| Elongation % | – | – | 13 | 9 | 4 | 4 |
| Expected 0.2% proof stress, min., MPa (tonf in$^{-2}$) | 140–170 (9.0–11.0) HB 100–150 | 230–260 (14.9–16.8) HB 100–130 | 165–200 (10.7–12.9) | 200–240 (12.5–15.5) | 360(23.3) | 260(16.8) |

*Not included in BS 1490: 1988.
† [ ] signifies obsolete specification.
‡ Si + Fe 1.0 max.
§ Or 5 days ageing at room temp.
** Can substitute stabilizing treatment at 200–250°C if used for pistons.
†† Allow to cool to 480°C before quench.

*Note:*

E – Excellent. F – Fair. G – Good. P – Poor. U – Unsuitable.
(Ge – General purpose alloy; SP – Special purpose alloy as per BS 1490: 1988).

**Table 4.4**  MISCELLANEOUS ALUMINIUM ALLOYS

| | | | | |
|---|---|---|---|---|
| *Specification* BS 1400; 1988 | LM5M(SP) | – | LM10TB(SP) | – |
| *Aerospace* | | | | |
| *BSL series* | – | – | 4L53 | L99 |
| *DTD series* | – | 5018A | – | – |

| *Composition %* (Single figures indicate maximum) | | | | |
|---|---|---|---|---|
| Copper | 0.1 | 0.2 | 0.1 | 0.1 |
| Magnesium | 3.0–6.0 | 7.4–7.9 | 9.5–11.0 | 0.20–0.45 |
| Silicon | 0.3 | 0.25 | 0.25 | 6.5–7.5 |
| Iron | 0.6 | 0.35 | 0.35 | 0.20 |
| Manganese | 0.3–0.7 | 0.1–0.3 | 0.10 | 0.10 |
| Nickel | 0.1 | 0.1 | 0.10 | 0.10 |
| Zinc | 0.1 | 0.9–1.4 | 0.10 | 0.10 |
| Lead | 0.05 | 0.05 | 0.05 | 0.05 |
| Tin | 0.1 | 0.05 | 0.05 | 0.05 |
| Titanium | 0.2 | 0.25 | 0.2[†] | 0.20 |
| Other | – | – | – | – |

| *Properties of material* | | | | |
|---|---|---|---|---|
| Suitability for: | | | | |
| Sand casting | F | F | F | G |
| Chill casting (gravity die) | F | F | F | E |
| Die casting (press die) | F[‡] | – | F[‡] | F[‡] |
| Strength at elevated temp. | F | F | F | – |
| Corrosion resistance | E | E | E | E |
| Pressure tightness | P | P | P | G |
| Fluidity | F | F | F | G |
| Resistance to hot shortness | F | G | G | G |
| Machinability | G | G | G | F |
| Melting range, °C | 580–642 | – | 450–620 | 550–615 |
| Casting temperature range, °C | 680–740 | 680–720 | 680–720 | 680–740 |
| Specific gravity | 2.65 | 2.64 | 2.57 | 2.67 |

| *Heat treatment* | | | | |
|---|---|---|---|---|
| Solution temperature, °C | – | 425–435[§] | 425–435 | 535–545 |
| Solution time, h | – | 8 | 8 | 12 |
| Quench | – | Oil at 160 °C** or boiling water | Oil at no more[††] than 160 °C | Water at 65 °C min |
| Precipitation temp., °C | – | – | – | 150–160 |
| Precipitation time, h | – | – | – | 4 |

| *Special properties* | Good corrosion resistance in marine atmospheres | – | Good shock resistance and high corrosion resistance *** | Excellent castability with good mech. props. |
|---|---|---|---|---|

| *Mechanical properties – Sand cast* – SI units (Imperial units in brackets) | | | | |
|---|---|---|---|---|
| Tensile stress min, MPa (tonf in$^{-2}$) | 140(9.1) | 278 | 280(18.0) | 230(14.9) |
| Elongation % | 3 | 3 | 8 | 2 |
| Expected 0.2% proof stress, min MPa (tonf in$^{-2}$) | 90–110(5.8–7.1) | 170(11.0) | 170–190 (11.0-12.3) | 185(12.0) |

| *Mechanical properties – chill cast* – SI units (Imperial units in brackets) | | | | |
|---|---|---|---|---|
| Tensile stress min, MPa (tonf in$^{-2}$) | 170(11.0) | 309(20.0) | 310(20.1) | 280(18.1) |
| Elongation % | 5 | 10 | 12 | 5 |
| Expected 0.2% proof stress, min MPa (tonf in$^{-2}$) | 90–120(5.8–7.8) | 170(11.0) | 170–200 (11.0–12.9) | 200(12.9) |

*[ ] obsolete.

[†]0.05% min. if Ti alone used for grain refinement.

[††]Not normally used in this form.

[§]Or 8 h at 435–445 C then raise to 490–500 C for further 8 h and quench as in table.

**Do not retain castings in oil for more than 1 h.

***Not generally recommended since occasional brittleness can develop over long periods.

**Table 4.4**  *(continued)*

| Specification BS 1400: 1988 | LM28TE(SP) | LM28TF(SP) | [LM23P]* | [LM15WP]* | – |
|---|---|---|---|---|---|
| Aerospace | | | | | |
| BSL series | – | – | 3L51 | 3L52 | |
| DTD series | – | – | – | – | 5008B |

*Composition % (Single figures indicate maximum)*

| | | | | | |
|---|---|---|---|---|---|
| Copper | | 1.3–1.8 | 0.8–2.0 | 1.3–3.0 | 0.1 |
| Magnesium | | 0.8–1.5 | 0.05–0.2 | 0.5–1.7 | 0.5–0.75 |
| Silicon | | 17–20 | 1.5–2.8 | 0.6–2.0 | 0.25 |
| Iron | | 0.7 | 0.8–1.4 | 0.8–1.4 | 0.5 |
| Manganese | | 0.6 | 0.1 | 0.1 | 0.1 |
| Nickel | | 0.8–1.5 | 0.8–1.7 | 0.5–2.0 | 0.1 |
| Zinc | | 0.2 | 0.1 | 0.1 | 4.8–5.7 |
| Lead | | 0.1 | 0.05 | 0.05 | 0.1 |
| Tin | | 0.1 | 0.05 | 0.05 | 0.05 |
| Titanium | | 0.2 | 0.25 | 0.25 | 0.15–0.25 |
| Other | | Cr 0.6 Co 0.5 | – | – | Cr 0.4–0.6 |

*Properties of material*
Suitability for:

| | | | | | |
|---|---|---|---|---|---|
| Sand casting | | P | G | F | F |
| Chill casting (gravity die) | | F | G | G | P |
| Die casting (press die) | | – | G‡ | U | U‡ |
| Strength at elevated temp. | | F | G | E | F |
| Corrosion resistance | | G | G | G | E |
| Pressure tightness | | F | G | F | F |
| Fluidity | | F | F | F | F |
| Resistance to hot shortness | | G | G | G | P |
| Machinability | | P | G | G | G |
| Melting range, °C | | 520–675 | 545–635 | 600–645 | 572–615 |
| Casting temp. range, °C | | ≮735 | 680–750 | 685–755 | 730–770 |
| Specific gravity | | 2.68 | 2.77 | 2.75 | 2.81 |

*Heat treatment*

| | | | | | |
|---|---|---|---|---|---|
| Solution temperature, °C | – | 495–505 | – | 520–540 | – |
| Solution time, h | – | 4 | – | 4 | – |
| Quench | – | Air blast | – | Water at 80–100 °C Oil or air blast | – |
| Precipitation temp, °C | 185 | 185 | 150–175 | 150–180 (195–205) | 175–185‡‡ (at least 24 h after cast) |
| Precipitation time, h | To produce required HB | 8 | 8–24 | 8–24(2–5) | 10(at least 24 h after cast) |
| Special properties | | Piston alloy | Aircraft engine castings | High mechanical props. at elevated temps. | Good strength without heat treatment. *See*‡‡ |

*Mechanical properties – sand cast* – SI units (Imperial units in brackets)

| | | | | | |
|---|---|---|---|---|---|
| Tensile stress min, MPa (tonf in⁻²) | – | 120(7.8) | 160(10.4) | 280(18.1) | 216(14.0) |
| Elongation % | – | – | 2 | – | 4 |
| Expected 0.2% proof stress, min MPa (tonf in⁻²) | – | – HB100–140 | 125(8.1) | 245(15.9) | 150(9.7) |

*Mechanical properties – chill cast* – SI units (Imperial units in brackets)

| | | | | | |
|---|---|---|---|---|---|
| Tensile stress min, MPa (tonf in⁻²) | 170(11.0) | 190(12.3) | 200(13.0) | 325(21.0) | 232(15.0) |
| Elongation % | – | – | 3 | – | 5 |
| Expected 0.2% proof stress min, MPa (tonf in⁻²) | – HB 90–130 | 160–190 (10.4–12.3) HB 100–140 | 140(19.1) | 295(19.1) | 180(11.7) |

†† Can be furnace cooled to 385–395 °C before quench. Do not retain in oil for more than 1 h. Further quench in water or air.
‡‡ Alternative-room temp. age-harden for 3 weeks.

*Note:*
E – Excellent. F – Fair. G – Good. P – Poor. U – Unsuitable.
(Ge – General purpose alloy; SP – Special purpose alloy as per BS 1490: 1988).

**Table 4.5**  HIGH STRENGTH CAST AL ALLOYS BASED ON AL-4.5 CU

| European Designation | KO1 | | | | A-U5GT | | | | | |
|---|---|---|---|---|---|---|---|---|---|---|
| Al Assoc (USA) designation | 201.0 | 201.2 | A201.0 | A201.2 | 204.0 | 204.2 | 206.0 | 206.2 | A206.0 | A206.2 |
| Cu | 4.0–5.2 | 4.0–5.2 | 4.0–5.0 | 4.0–5.0 | 4.2–5.0 | 4.2–4.9 | 4.2–5.0 | 4.2–5.0 | 4.2–5.0 | 4.2–5.0 |
| Mg | 0.15–0.55 | 0.20–0.55 | 0.15–0.35 | 0.20–0.35 | 0.15–0.35 | 0.20–0.35 | 0.15–0.35 | 0.20–0.35 | 0.15–0.35 | 0.20–0.35 |
| Si | 0.10 | 0.10 | 0.05 | 0.05 | 0.20 | 0.15 | 0.10 | 0.10 | 0.05 | 0.05 |
| Fe | 0.15 | 0.10 | 0.10 | 0.07 | 0.35 | 0.10–0.20 | 0.15 | 0.10 | 0.10 | 0.07 |
| Mn | 0.20–0.50 | 0.20–0.50 | 0.20–0.40 | 0.20–0.40 | 0.10 | 0.05 | 0.20–0.50 | 0.20–0.50 | 0.20–0.50 | 0.20–0.50 |
| Ni | – | – | – | – | 0.05 | 0.03 | 0.05 | 0.03 | 0.05 | 0.03 |
| Zn | – | – | – | – | 0.10 | 0.05 | 0.10 | 0.05 | 0.10 | 0.05 |
| Sn | – | – | – | – | 0.05 | 0.05 | 0.05 | 0.05 | 0.05 | 0.05 |
| Ti | 0.15–0.35 | 0.15–0.35 | 0.15–0.35 | 0.15–0.35 | 0.15–0.30 | 0.15–0.25 | 0.15–0.3 | 0.15–0.25 | 0.15–0.30 | 0.15–0.25 |
| Ag | 0.40–1.0 | 0.40–1.0 | 0.40–1.0 | 0.40–1.0 | – | – | – | – | – | – |
| Others each | 0.05 | 0.05 | 0.03 | 0.03 | 0.05 | 0.05 | 0.05 | 0.05 | 0.05 | 0.05 |
| Others total | 0.10 | 0.10 | 0.10 | 0.10 | 0.15 | 0.15 | 0.15 | 0.15 | 0.15 | 0.15 |

**Table 4.6**  MINIMUM REQUIREMENTS FOR SEPARATELY CAST TEST BARS

| Alloy | Sand/Die | Treatment* | UTS MPa | 0.2 PS MPa | El% in 50 mm or 4 × diam | Typical HB 500 kgf 10 mm |
|---|---|---|---|---|---|---|
| 201.0 | Sand | T6 | 414 | 345 | 5.0 | 110–140 |
|  | Sand | T7 | 414 | 345 | 3.0 |  |
| 204.0 | Sand | T4 | 311 | 194 | 6.0 | 130 |
|  | Die | T4 | 331 | 220 | 8.0 | 95 |

*For temper designations see Table 7.3.

**Table 4.7**  TYPICAL PROPERTIES OF SEPARATELY CAST TEST BARS

| Alloy | Sand/Die | Treatment* | | | UTS MPa | 0.2 PS MPa | El% in 50 mm or 4 × diam | Typical HB 500 kgf 10 mm |
|---|---|---|---|---|---|---|---|---|
| 201.0 | Sand | T4 | | | 365 | 215 | 20 | 95 |
|  |  | T43 | | | 414 | 255 | 17 |  |
|  | Sand | T6 | | | 448–485 | 380–435 | 7–8 | 135 |
|  | Sand | T7 | Room | Temp. | 460–469 | 415 | 4.5–5.5 | 130 |
|  |  | T7 | 150C* | 0.5–100 h | 380 | 360 | 6–8.5 |  |
|  |  |  |  | 1 000 h | 360 | 345 | 8 |  |
|  |  |  |  | 10 000 h | 315 | 275 | 7 |  |
|  |  |  | 205C | 0.5 h | 325 | 310 | 9 |  |
|  |  |  |  | 100 h | 285 | 270 | 10 |  |
|  |  |  |  | 1 000 h | 250 | 230 | 9 |  |
|  |  |  |  | 10 000 h | 185 | 150 | 14 |  |
|  |  |  | 260C | 0.5 h | 195 | 185 | 14 |  |
|  |  |  |  | 100 h | 150 | 140 | 17 |  |
|  |  |  |  | 1 000 h | 125 | 110 | 18 |  |
|  |  |  | 315C | 0.5 h | 140 | 130 | 12 |  |
|  |  |  |  | 100 h | 85 | 75 | 30 |  |
|  |  |  |  | 1 000 h | 70 | 60 | 39 |  |
|  |  |  |  | 10 000 h | 60 | 55 | 43 |  |
| A206.0 | Sand | T4 | | | | | | 118HV |
|  |  | T7 | | | | | | 137HV |
|  |  |  | 20C |  | 436 | 347 | 11.7 |  |
|  |  |  | 120C |  | 384 | 316 | 14.0 |  |
|  |  |  | 175C |  | 333 | 302 | 17.7 |  |

*Elevated temperature properties.

## 4.2 Magnesium alloys

**Table 4.8** ZIRCONIUM-FREE MAGNESIUM ALLOYS

*Grain refined (0.05–0.2 mm chill cast) when superheated to 850–900 °C or suitably treated with carbon (as hexachlorethane)*

| *Elektron designation* | A8 | | A8 (High purity) | |
| *ASTM designation* | AZ81 | | AZ81 | |
| --- | --- | --- | --- | --- |
| | MAG1M* (GP)† | MAG1TB* (GP) | MAG2M(SP)† | MAG2TB(SP) |
| *Specifications BS 2970: 1989* | | | | |
| *BSS L series* | – | 3L. 112 | – | – |
| *Equivalent DTD* | – | – | 684A | 690A |
| *Composition %* (Single figures indicate maximum) | | | | |
| Aluminium | 7.5–9.0 | | 7.5–9.0 | |
| Zinc | 0.3–1.0 | | 0.3–1.0 | |
| Manganese | 0.15–0.4 | | 0.15–0.7 | |
| Copper | 0.15 | | 0.005 | |
| Silicon | 0.3 | | 0.01 | |
| Iron | 0.05 | | 0.003 | |
| Nickel | 0.01 | | 0.001 | |
| Cu+Si+Fe+4Ni | 0.40 | | | |
| *Material properties* | | | | |
| Founding | Good | | Good | |
| Characteristics | Sand and permanent‡ mould | | Special melting technique required | |
| Tendency to hot tearing | Little | | Little | |
| Tendency to micro-porosity | Appreciable | | Appreciable | |
| Castability§ | A | | A | |
| Weldability (Ar-Arc process) | Good | | Good | |
| Relative damping capacity¶ | C | | C | |
| Strength at elevated temperature** | C | | C | |
| Corrosion resistance | Moderate | | Moderate | |
| Density, g cm⁻³ | 1.81 | | 1.81 | |
| Liquids, °C | 600 | | 600 | |
| Solidus, °C | 475 | | 475 | |
| Non-equilibrium solidus, °C | 420 | | 420 | |
| Casting temperature range, °C | 680–800 | | 680–800 | |

| | MAG3M(GP) | MAG3TB(GP) | MAG3TF(GP) | MAG7M(GP) | MAG7TF(GP) |
|---|---|---|---|---|---|
| *Heat treatment* | | | | | |
| Solution: | | | | | |
| Time, h | – | 12 (min) | – | – | 12 (min) |
| Temperature, °C | – | 435 (max)‡‡ | – | – | 435 (max)‡‡ |
| Cooling | – | Air, oil or water | – | – | Air, oil or water |
| *Elektron designation* | | AZ91 | | | *C alloy* |
| *ASTM designation* | | AZ91 | | | |
| *Specifications* BS 2970: 1989 | MAG3M(GP) | MAG3TB(GP) | MAG3TF(GP) | MAG7M(GP) | MAG7TF(GP) |
| *BSS L series* | – | 3L.124 | 3L.125 | – | – |
| *Equivalent DTD* | – | – | – | – | – |
| *Composition %* (Single figures indicate maximum) | | | | | |
| Aluminium | | 9.0–10.5 | | | 7.5–9.5 |
| Zinc | | 0.3–1.0 | | | 0.3–1.5 |
| Manganese | | 0.15–0.4 | | | 0.15–0.8 |
| Copper | | 0.15 | | | 0.35 |
| Silicon | | 0.3 | | | 0.40 |
| Iron | | 0.05 | | | 0.05 |
| Nickel | | 0.01 | | | 0.02 |
| Cu + Si + Fe + Ni | | 0.40 | | | 0.75 |
| *Material properties* | | | | | |
| Founding | | Good | | | Good |
| Characteristics | | Sand, permanent mould and die (pressure) | | | Sand, permanent mould and die (pressure) |
| Tendency to hot tearing | | Little | | | Little |
| Tendency to micro-porosity | | Less than MAG1 | | | Less than MAG1 |
| Castability§ | | A | | | A |
| Weldability (Ar-Arc process) | | Good, but some difficulty with die castings | | | Good, but some difficulty with die castings |
| Relative damping capacity¶ | | C | | | C |
| Strength at elevated temperature** | | C | | | C |
| Corrosion resistance | | Moderate | | | Moderate |
| Density, g cm$^{-3}$ | | 1.83 | | | 1.82 |
| Liquidus, °C | | 595 | | | 600 |
| Solidus, °C | | 470 | | | 475 |
| Non-equilibrium solidus, °C | | 420 | | | 420 |
| Casting temperature range, °C | | 680–800 | | | 680–800 |

*continued overleaf*

**Table 4.8**  (*continued*)

| | | | |
|---|---|---|---|
| *Heat treatment* | | | |
| *Solution:* | | | |
| Time, h | — | 16 (min) | 16 (min) |
| Temperature, °C | — | 435 (max)‡‡ | 435 (max)‡‡ |
| Cooling | — | Air, oil or water | Air, oil or water |
| | | | |
| *Elektron designation* | AZ91E | AZ91 (HP) | ZC63 |
| *ASTM designation* | AZ91E | AZ91D | ZC63 |
| | | | |
| *Specifications* BS 2970: 1989 | | MAG11 (GP) | — |
| *BSS L series* | | — | — |
| *Equivalent DTD* | | — | — |
| | | | |
| *Composition* % (Single figures indicate maximum) | | | |
| Aluminium | | 8.5–9.5 | — |
| Zinc | | 0.45–0.9 | 5.5–6.5 |
| Manganese | | 0.15–0.40 | 0.25–0.75 |
| Copper | | 0.015 | 2.4–3.0 |
| Silicon | | 0.020 | 0.20 |
| Iron | | 0.005 | 0.05 |
| Nickel | | 0.0010 | 0.01 |
| Cu + Si + Fe + Ni | | — | — |
| | | | |
| *Material properties* | | | |
| Founding | Sand and permanent mould | High pressure die | Sand, permanent and high pressure die |
| Characteristics | Good | Good | Good |
| Tendency to hot tearing | Little | Little | Little |
| Tendency to micro-porosity | Less than MAG1 | Little | Little |
| Castability§ | A | A | B |
| Weldability (Ar-Arc process) | Good | Difficult | Good |
| Relative damping capacity¶ | C | C | C |
| Strength at elevated temperature** | C | C | B |
| Corrosion resistance | Excellent | Excellent | Moderate |
| Density, g cm$^{-3}$ | 1.83 | 1.83 | 1.84 |
| Liquidus, °C | 595 | 595 | 635 |
| Solidus, °C | 470 | 470 | 465 |
| Non-equilibrium solidus, °C | 420 | 420 | — |
| Casting temperature range, °C | 680–800 | 620–680 | 700–810 |

| | MAG1M* (GP)† | MAG1TB* (GP) | MAG2M(SP)† | MAG2TB(SP) |
|---|---|---|---|---|
| **Heat treatment** | | | | |
| Time, h | | 16 (min) | | 8 |
| Temperature, °C | Not suitable | 435 (max) | | 440 (max) |
| Cooling | | Air, oil or water | | oil or water |
| **Elektron designation** | | A8 | | A8 (High purity) |
| **ASTM designation** | | AZ81 | | AZ81 |
| **Specifications** BS 2970: 1989 | MAG1M* (GP)† | MAG1TB* (GP) | MAG2M(SP)† | MAG2TB(SP) |
| BSS L series | – | 3L.122 | – | – |
| Equivalent DTD | – | – | 684A | 690A |
| **Heat treatment – continued** | | | | |
| Precipitation: | | | | |
| Time, h | – | – | – | – |
| Temperature, °C | – | – | – | – |
| Stress relief: | | | | |
| Time, h | 2–4 | – | 2–4 | – |
| Temperature, °C | 250–330 | – | 250–330 | – |
| *Mechanical properties – sand cast* – (SI units first, Imperial units following in brackets) | | | | |
| Tensile strength (min), MPa (tonf in$^{-2}$) | 140 (9.1) | 200 (13.0) | 140 (9.1) | 200 (13.0) |
| 0.2% proof stress (min), MPa (tonf in$^{-2}$) | 85 (5.5) | 80 (5.2) | 85 (5.5) | 80 (5.2) |
| Elongation % (min) (5.65$\sqrt{S_0}$) | 2 | 6 | 2 | 6 |
| *Mechanical properties – chill cast* – (SI units first, Imperial units following in brackets) | | | | |
| Tensile strength (min), MPa (tonf in$^{-2}$) | 185 (12.0) | 230 (14.9) | 185 (12.0) | 230 (14.9) |
| 0.2% proof stress (min), MPa (tonf in$^{-2}$) | 85 (5.5) | 80 (5.2) | 85 (5.5) | 80 (5.2) |
| Elongation % (min) (5.65$\sqrt{S_0}$) | 4 | 10 | 4 | 10 |
| Applications | Automobile road wheels | Good ductility and shock resistance | High-purity alloy – offers good corrosion resistance | |

*continued overleaf*

**Table 4.8** (*continued*)

| Elektron designation ASTM designation | AZ91 | | | C alloy | | |
|---|---|---|---|---|---|---|
| | MAG3M(GP) | MAG3TB(GP) | MAG3TF(GP) | MAG7M(GP) | MAG7TB(GP) | MAG7TF(GP) |
| *Specifications* BS 2970: 1989 | | 3L.124 | 3L.125 | | | |
| *BSS L series* | – | | | – | – | – |
| *Equivalent DTD* | – | – | – | – | – | – |
| Precipitation: | | | | | | |
| Time, h | – | – | 8(min) | – | – | 8(min) |
| Temperature, °C | – | – | 210(max) | – | – | 210(max) |
| Stress relief: | | | | | | |
| Time, h | 2–4 | – | – | 2–4 | – | – |
| Temperature, °C | 250–330 | – | – | 250–330 | – | – |
| *Mechanical properties – sand cast* – (SI units first, Imperial units following in brackets) | | | | | | |
| Tensile strength (min), MPa (tonf in$^{-2}$)†† | 125 (8.1) | 200 (13.0) | 200 (13.0) | 125 (8.1) | 185 (12.0) | 185 (12.0) |
| 0.2% proof stress (min), MPa (tonf in$^{-2}$) | 95 (6.2) | 85 (5.5) | 130 (8.4) | 85 (5.5) | 80 (5.2) | 110 (7.1) |
| Elongation % (min) (5.65√$S_0$) | – | 4 | – | – | 4 | – |
| *Mechanical properties – chill cast* – (SI units first, Imperial units following in brackets) | | | | | | |
| Tensile strength (min), MPa (tonf in$^{-2}$) | 170 (11.0) | 215 (13.9) | 215 (13.9) | 170 (11.0) | 215 (13.9) | 215 (13.9) |
| 0.2% proof stress (min), MPa (tonf in$^{-2}$) | 100 (6.5) | 85 (5.5) | 130 (8.4) | 85 (5.5) | 80 (5.2) | 110 (7.1) |
| Elongation % (min) (5.66√$S_0$) | 2 | 5 | 2 | 2 | 5 | 2 |
| Applications | For pressure tight applications | | Increased proof stress after full heat treatment | Principal alloy for commercial usage | | |

| | AZ91E | AZ91(HP) | AZ91D | ZC63 |
|---|---|---|---|---|
| *Elektron designation* | | | | |
| ASTM designation | AZ91E | AZ91(HP) | AZ91D | ZC63 |
| *Specifications* BS 2970: 1989 | | MAG11(GP) | | ZC63 |
| BSS L series | | | – | – |
| Equivalent DTD | | – | – | – |
| Precipitation: | | | | |
| Time, h | 8 (min) | | Not suitable | 16 (min) |
| Temperature, °C | 210 (max) | | | 200 (max) |
| Stress relief: | | | | |
| Time, h | – | | – | – |
| Temperature, °C | – | | – | – |
| *Mechanical properties – sand cast* – (SI units first, Imperial units following in brackets) | | | | |
| Tensile strength (min), MPa (tonf in$^{-2}$) | 200 | | Typical high pressure die-cast properties | 210 |
| 0.2% proof stress (min), MPa (tonf in$^{-2}$) | 130 | | | 125 |
| Elongation % (min) $(5.65\sqrt{S_0})$ | 2 | | | 3 |
| *Mechanical properties – chill cast* – (SI units first, Imperial units following in brackets) | | | | |
| Tensile strength (min), MPa (tonf in$^{-2}$) | 215 | | 200 | 210 |
| 0.2% proof stress (min), MPa (tonf in$^{-2}$) | 130 | | 150 | 125 |
| Elongation % (min) $(5.65\sqrt{S_0})$ | 2 | | 1 | 3 |
| *Applications* | | | High purity alloy – offers excellent corrosion resistance. Max. temp. 120 °C | Better foundability than AZ91 with superior elevated temperature properties |

*M-as cast.
TS – Stress relieved only.
TE – Precipitation treated only.
TB – Solution treated only.
TF – Solution and precipitation treated.
† GP General purpose alloy.
SP Special purpose alloy.
‡ Permanent mould=gravity die casting.
§ Ability to fill mould easily. A, B, C, indicate decreasing castability.

¶ Damping capacity ratings.
A=Outstanding; better than grey cast iron.
B=Equivalent to cast-iron.
C=Infersior to cast-iron but better Al-base cast alloys.
** A=Particularly recommended.
B=Suitable but not especially recommended.
C=Not recommended where strength at elev. temps is likely to be an important consideration.
†† 1 MPa=1 H m$^{-2}$ =0 064 75 tonf in$^{-2}$.
‡‡ SO$_2$ or CO$_2$ atmosphere.

**Table 4.9**  MAGNESIUM-ZIRCONIUM ALLOYS
*Inherently fine grained (0.015–0.035 mm chill cast)*

| | Z5Z<br>ZK51 | RZ5<br>ZE41 | ZRE1<br>EZ33 |
|---|---|---|---|
| *Elektron designation*<br>*ASTM designation* | | | |
| *Specifications* BS 2970: 1989<br>*BSS L series*<br>*Equivalent DTD* | MAG4TE* (GP)†<br>2L.127<br>– | MAG5TE(SP)<br>2L.128<br>– | MAG6TE(SP)<br>2L.126<br>– |
| *Composition % (Single figures indicate maximum)* | | | |
| Zinc | 3.5–5.5 | 3.5–5.5 | 0.8–3.0 |
| Silver | – | – | – |
| Rare earth metals | – | 0.75–1.75 | 2.5–4.0 |
| Thorium | – | – | – |
| Zirconium | 0.4–1.0 | 0.4–1.0 | 0.4–1.0 |
| Copper | 0.03 | 0.03 | 0.03 |
| Nickel | 0.005 | 0.005 | 0.005 |
| Iron | – | – | – |
| Silicon | – | – | – |
| Manganese | – | – | – |
| *Material properties* | | | |
| Founding characteristics | Good in sand and permanent moulds§ | Good in sand and permanent moulds | Excellent in sand and permanent moulds |
| Tendency to hot tearing | Marked | Some | Little |
| Tendency to micro-porosity | Very appreciable | Virtually none | None |
| Castability¶ | B | A | A |
| Weldability (Ar-Arc Process) | Not recommended | Moderate | Very good |
| Relative damping capacity** | B/C | B/C | B |
| Strength at elevated temperature†† | C | B | A |
| Resistance to creep at elevated temperature | Poor | Moderate | Good up to 250°C |
| Corrosion resistance | Moderate | Moderate | Moderate |
| Density, g cm$^{-3}$ (20°C) | 1.81 | 1.84 | 1.80 |
| Liquidus, °C | 640 | 640 | 640 |
| Solidus, °C | 560 | 510 | 545 |
| Casting temperature range, °C | 720–810 | 720–810 | 720–810 |

| | | | |
|---|---|---|---|
| *Heat treatment* | | | |
| Solution: | | | |
| Time, h | 16 | — | 8 |
| Temperature, °C | 180 | — | 200 |
| Cooling | Air cool | — | Air cool |
| Precipitation: | | | |
| Time, h | 2 | 2 followed by 16 | 10 |
| Temperature, °C | 330 | 330    180 | 250 max |
| Cooling | | Air cool after each | Air cool |
| Post – weld stress relief: | | | |
| Time, h | to precede precipitation treatment | Precipitation treatment affords s/relief | — |
| *Mechanical properties – sand cost –* SI units (Imperial units in brackets) | | | |
| Tensile strength min, MPa (tonf in$^{-2}$) | 230 (14.9) | 200 (13.0) | 140 (9.1) |
| 0.2% proof stress min, MPa (tonf in$^{-2}$) | 145 (9.4) | 135 (8.7) | 95 (6.2) |
| Elongation, % ($5.65\sqrt{S_0}$) min | 5 | 3 | 3 |
| *Mechanical properties – chill cast –* SI units (Imperial units in brackets) | | | |
| Tensile strength min, MPa (tonf in$^{-2}$) | 245 (15.9) | 215 (13.9) | 155 (10.0) |
| 0.2% proof stress min, MPa (tonf in$^{-2}$) | 145 (9.4) | 135 (8.7) | 110 (7.1) |
| Elongation, % ($5.65\sqrt{S_0}$) min | 7 | 4 | 3 |
| Applications | High strength plus good ductility. Not suitable for spidery complex shapes | For high-strength pressure-tight applications | High degree of pressure tightness at room and elevated temperatures |

*continued overleaf*

**Table 4.9** (*continued*)

| | ZT1‡‡‡ | TZ6‡‡‡ | ZE63 |
|---|---|---|---|
| *Elektron designation* | ZT1‡‡‡ | TZ6‡‡‡ | ZE63 |
| *ASTM designation* | HZ32 | ZH62 | ZE63 |
| *Specifications* BS 2970: 1989 | MAG8TE(SP) | MAG9TE(SP) | |
| *BSS L series* | 5005A | 5015A | — |
| *Equivalent DTD* | — | — | 5045 |
| *Composition % (Single figures indicate maximum)* | | | |
| Zinc | 1.7–2.5 | 5.0–6.0 | 5.5–6.0 |
| Silver | 0.10 | 0.20 | |
| Rare earth metals | | | 2.0–3.0 |
| Thorium | 2.5–4.0 | 1.5–2.3 | — |
| Zirconium | 0.4–1.0 | 0.4–1.0 | 0.4–1.0 |
| Copper | 0.03 | 0.03 | 0.03 |
| Nickel | 0.005 | 0.005 | 0.005 |
| Iron | 0.01 | 0.01 | 0.01 |
| Silicon | 0.01 | 0.01 | 0.01 |
| Manganese | 0.15 | 0.15 | 0.15 |
| *Material properties* | | | |
| Founding characteristics | As per MAG7 but more sluggish | Similar to MAG5 | Good |
| Tendency to hot tearing | Little | Very little | Negligible |
| Tendency to micro-porosity | None | Low | Virtually none |
| Castability¶ | C | B | A |
| Weldability (Ar-Arc process) | Very good | Fair | Very good*** |
| Relative damping capacity** | B | C | B/C |
| Strength at elevated temperature†† | A | B | C |
| Resistance to creep at elevated temperature | Good up to 350°C | Fair | Poor |
| Corrosion resistance | Moderate | Moderate | Moderate |
| Density, g cm⁻³ (20°) | 1.85 | 1.87 | 1.87 |
| Liquidus, °C | 645 | 630 | 625 |
| Solidus, °C | 550 | 520 | 516 |
| Casting temperature range, °C | 720–810 | 720–810 | 720–810 |

| *Heat treatment* | | | |
|---|---|---|---|
| Solution: | | | |
| Time, h | 16 | — | 30 for 12 mm sctn. 70 for 25 mm sctn. |
| Temperature, °C | 315 | — | 480††† |
| Cooling | Air cool | — | Air blast or water spray |
| Precipitation: | | | |
| Time, h | 2 | 2 followed 16 | 48 or 72 |
| Temperature, °C | 350 | 330 by 180 | 138 127 |
| Cooling | Air cool | Air cool after each | Air cool |
| Post weld stress relief: | | | |
| Time, h | | Pptn. treatment affords stress | — |
| Temperature, °C | | relief | — |
| *Mechanical properties – sand cast – SI units (Imperial units in brackets)* | | | |
| Tensile strength min, MPa (tonf $in^{-2}$) | 185 (12.0) | 255 (16.5) | 275 (17.8) |
| 0.2% proof stress min, MPa (tonf $in^{-2}$) | 85 (5.5) | 155 (10.0) | 170 (11.0) |
| Elongation, % (5.65√$S_0$) min | 5 | 5 | 5 |
| *Mechanical properties – chill cast – SI units (Imperial units in brackets)* | | | |
| Tensile strength min, MPa (tonf $in^{-2}$) | 185 (12.0) | 255 (16.5) | |
| 0.2% proof stress min, MPa | 85 (5.5) | 155 (10.0) | |
| Elongation, % (5.65√$S_0$) min | 5 | 5 | |
| Applications | Creep resistant alloy | For heavy duty structural usage | Sand Cast Alloy High strength with good ductility and excellent fatigue resistance. Structural parts aircraft, etc. |

*continued overleaf*

**Table 4.9**  (*continued*)

| Elektron designation ASTM | MSR-A | MSR-B | MSR QE22 | MTZ‡‡‡ HK31 | EQ21 EQ21 | WE54 WE54 | WE43 WE43 |
|---|---|---|---|---|---|---|---|
| Specifications BS 2970: 1972 | – | MAG12TF(SP) | – | – | MAG13TF(SP) | MAG14TF(SP) | – |
| BSS L series | | | | | | | |
| Equivalent DTD | 5025A | 5035A | 5055 | – | – | – | – |
| *Composition %* (Single figures indicate maximum) | | | | | | | |
| Zinc | 0.2 | 0.2 | 0.2 | 0.3 | 0.2 | 0.2 | 0.2 |
| Silver | 2.0–3.0 | 2.0–3.0 | 2.0–3.0 | – | 1.3–1.7 | – | – |
| Rare earth metals | 1.2–2.0‡ | 2.0–3.0‡ | 1.8–2.5‡ | 0.1 | 1.5–3.0‡ | 2.0–4.0¶¶¶ | 2.4–4.4¶¶¶ |
| Thorium | – | – | – | 2.5–4.0 | – | – | – |
| Zirconium | 0.4–1.0 | 0.4–1.0 | 0.4–1.0 | 0.4–1.0 | 0.4–1.0 | 0.4–1.0 | 0.4–1.0 |
| Copper | 0.03 | 0.03 | 0.03 | 0.03 | 0.05–0.10 | 0.03 | 0.03 |
| Nickel | 0.005 | 0.005 | 0.005 | 0.005 | 0.005 | 0.005 | 0.005 |
| Iron | 0.01 | 0.01 | 0.01 | 0.01 | 0.01 | 0.01 | 0.01 |
| Silicon | 0.01 | 0.01 | 0.01 | 0.01 | 0.01 | 0.01 | 0.01 |
| Manganese | 0.15 | 0.15 | 0.15 | 0.15 | 0.15 | 0.15 | 0.15 |
| Yttrium | – | – | – | – | – | 4.75–5.5 | 3.7–4.3 |

| | MSR-A | MSR-B | MSR | MTZ‡‡‡ | EQ21 | WE54 | WE43 |
|---|---|---|---|---|---|---|---|
| *Elektron designation* | MSR-A | MSR-B | MSR | MTZ‡‡‡ | EQ21 | WE54 | WE43 |
| *ASTM designation* | – | – | QE22 | MK31 | EQ21 | WE54 | WE43 |
| *Specifications* BS 2970: 1989 | – | MAG12TF (SP) | – | – | MAG13TF(SP) | MAG14TF(SP) | – |
| *BSS L series* *Equipment DTD* | 5025A | 5035A | 5055 | – | – | – | – |
| *Material properties* Founding characteristics | Good | Good | Good | Less easy to found than MSR types | Good | Good | Good |
| Tendency to hot tearing | Little | Little | Little | Very little | Little | Very little | Very little |
| Tendency to micro-porosity | Slight | Slight | Slight | Negligible | Slight | Slight | Slight |
| Castability¶ | B | B | B | C | B | B | B |
| Weldability (Ar-Arc process) | Very good | Very good | Very good | Very good | Very good | Very good | Very good |
| Relative damping capacity** | B/C | B/C | B/C | B/C | B/C | B/C | B/C |
| Strength at elevated temperature†† | A | A | A | A | A | A | A |
| Resistance to creep at elevated temperature | Good up to 200°C | Good up to 200°C | Good up to 200°C | Good up to 350°C for short time applications | Good up to 200°C | Very good up to 250°C | Very good up to 250°C |
| Corrosion resistance | Moderate | Moderate | Moderate | Moderate | Moderate | Excellent | Excellent |
| Density, g cm$^{-3}$ (20°C) | 1.81 | 1.82 | 1.81 | 1.84 | 1.81 | 1.85 | 1.85 |
| Liquidus, °C | 640 | 640 | 640 | 645 | 640 | 640 | 640 |
| Solidus, °C | 550 | 550 | 550 | 590 | 545 | 550 | 550 |
| Casting temperature range, °C | 720–810 | 720–810 | 720–810 | 720–810 | 720–810 | 720–810 | 720–810 |
| *Heat treatment* Solution: Time, h | 8 | 8 | 8 | 2 | 8 | 8 | 8 |
| Temperature, °C | 525‡‡ | 525‡‡ | 525‡‡ | 565‡‡ ¶¶ | 520‡‡ | 525‡‡ | 525‡‡ |
| Cooling | Water or oil | Water or oil | Water or oil | Air cool | Water or oil | Air cool | Water or oil |
| Precipitation: Time, h | 16 | 16 | 16 | 16 | 16 | 16 | 16 |
| Temperature, °C | 200 | 200 | 200 | 200 | 200 | 250 | 250 |
| | Air cool | Air cool | Air cool | Air cool | Air cool | Air cool | Air cool |

*continued overleaf*

**Table 4.9** (*continued*)

| | MSR-A | MSR-B | MSR | MTZ‡‡‡ | EQ21 | WE54 | WE43 |
|---|---|---|---|---|---|---|---|
| Elektron designation | MSR-A | MSR-B | MSR | MTZ‡‡‡ | EQ21 | WE54 | WE43 |
| ASTM designation | — | — | QE22 | MK31 | EQ21 | WE54 | WE43 |
| **Specifications BS 2970: 1989** | — | MAG12TF(SP) | — | — | MAG13TF(SP) | MAG14TF(SP) | MAG14TF(SP) |
| BSS L series | | | | | | | |
| Equipment DTD | 5025A | 5035A | 5055 | — | — | — | — |
| **Post-weld stress relief:** | | | | | | | |
| Time, h | 1 | | | Repeat above cycle | 1 | 1 | 1 |
| Temperature, °C | 510 followed by above quench and age | | | | 505 followed by above quench and age | 510 followed by above aircool and age | 510 followed by above quench and age |
| *Mechanical properties – sand cast* – SI units (Imperial units in brackets) | | | | | | | |
| Tensile strength min, MPa (tonf in⁻²) | 240 (15.5) | 240 (15.5) | 240 (15.5) | 200 (13.0) | 240 | 250 | 250 |
| 0.2% proof stress min, MPa (tonf in⁻²) | 170 (11.0) | 185 (12.0) | 175 (11.3) | 93 (6.0) | 170 | 175 | 165 |
| Elongation, % (5.65 √$S_0$) min | 4 | 2 | 2 | 5 | 2 | 2 | 2 |
| *Mechanical properties – chill cast* – SI units (Imperial units in brackets) | | | | | | | |
| Tensile strength min, MPa (tonf in⁻²) | 240 (15.5) | 240 (15.5) | 240 (15.5) | | 240 | 250 | 250 |
| 0.2% proof stress min, MPa (tonf in⁻²) | 170 (11.0) | 185 (12.1) | 175 (11.3) | Usually sand | 170 cast | 175 | 165 |
| Elongation, % (5.65 √$S_0$) min | 4 | 2 | 2 | 2 | 2 | 2 | 2 |
| Applications | High strength in thick and thin section castings. Good elevated temperature (up to 250 °C) short time tensile and fatigue props. | | Similar to MSRA-B | Superior short time tensile and creep resistance at temperatures around 300 °C | Similar to MSR alloys but less | Excellent strength up to 300 °C for short time applications. Excellent corrosion resistance | Excellent strength up to 250 °C for long time applications. Excellent corrosion resistance |

*See footnote to Table 4.1.
†See footnote to Table 4.1.
‡Neodymium-rich rare rearths (others Ce-rich).
¶See footnote to Table 4.1.
**See footnote to Table 4.1.
§See footnote to Table 4.1.

††*See* footnote to Table 4.1.
‡‡SO₂ or CO₂ atmosphere.
¶¶Castings to be loaded into furnace at operating temperature.
***But only before hydriding treatment.
†††In hydrogen at atmospheric pressure.
‡‡‡Thorium containing alloys are being replaced by alternative magnesium based alloys.
¶¶¶Neodymium and heavy rare earths.

# 5 Equilibrium diagrams

## 5.1 Index of binary diagrams

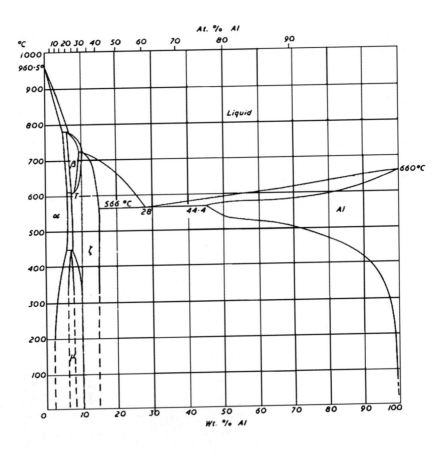

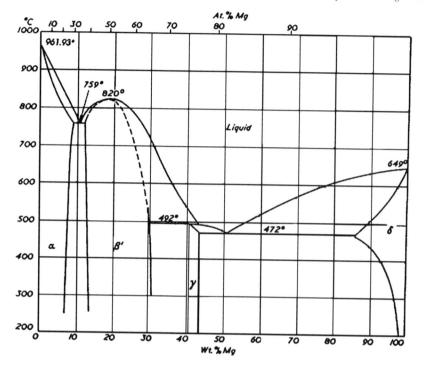

Ag–Mg

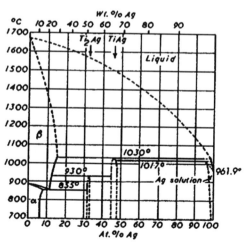

Ag–Ti

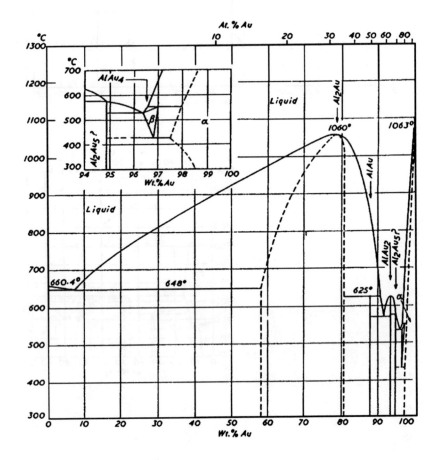

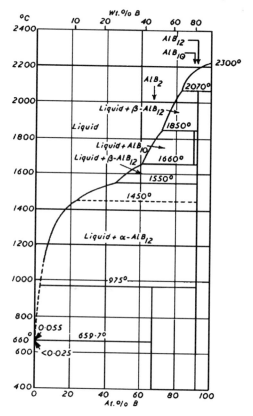

Al–B

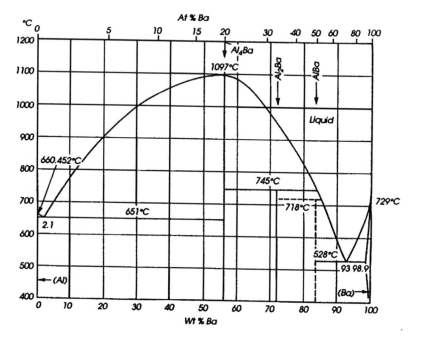

Al–Ba

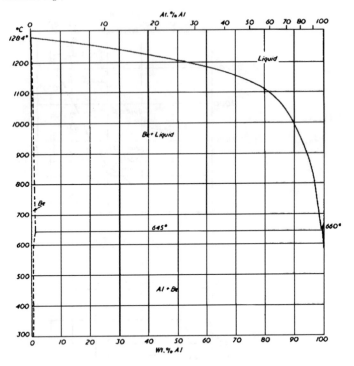

Al–Be

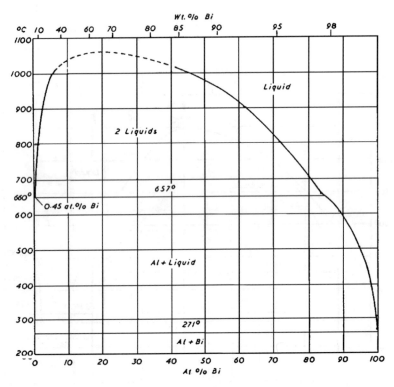

Al–Bi

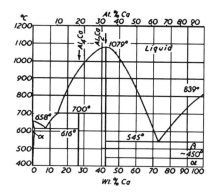

Al–Ca

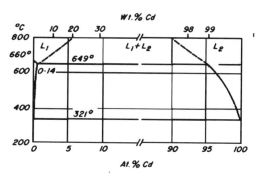

Al–Cd

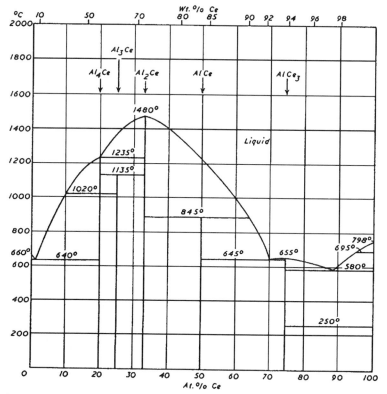

Al–Ce

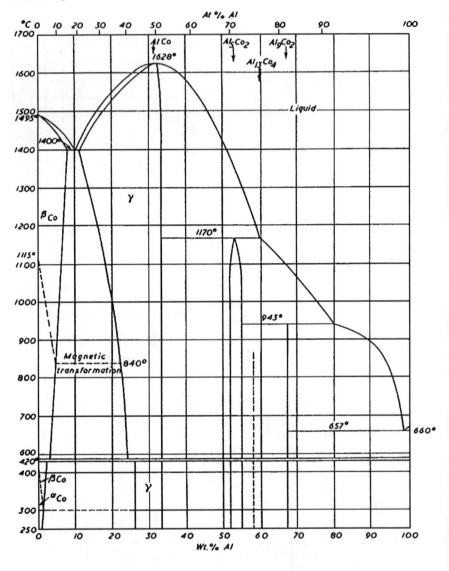

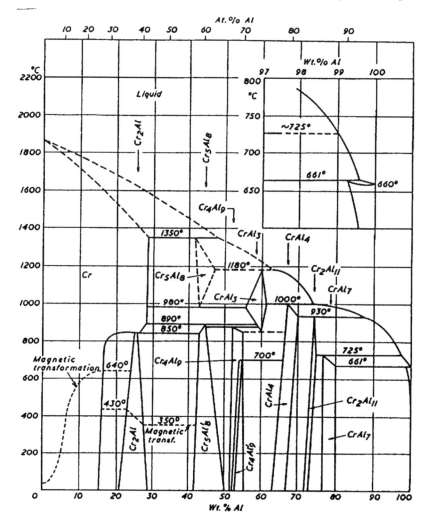

Al–Cr

**Al–Cu**

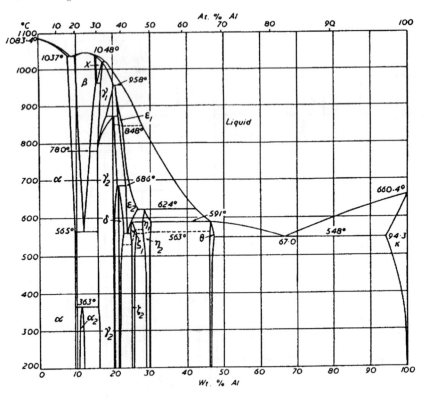

**Al–Cs**

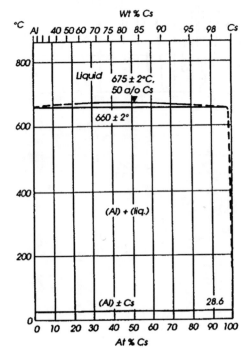

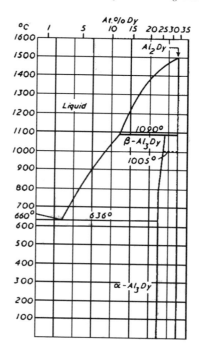

Al–Dy

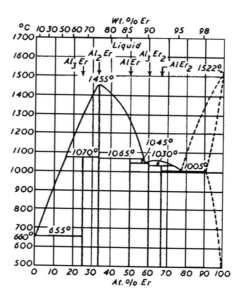

Al–Er

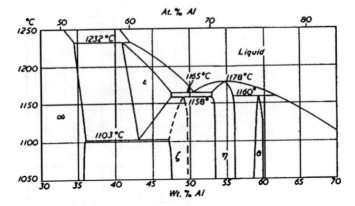

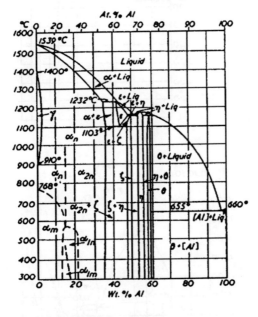

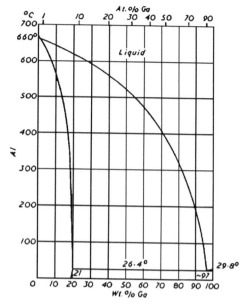

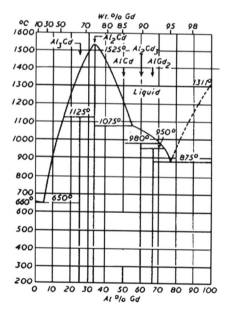

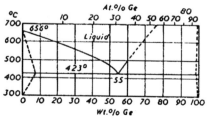

Al–Hf

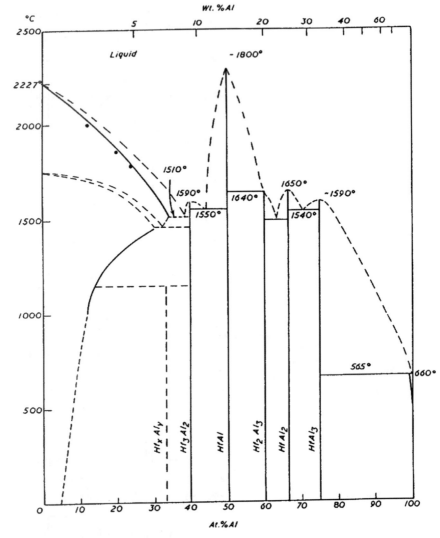

Al–Hg

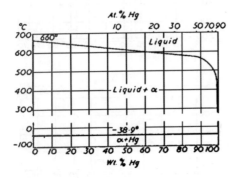

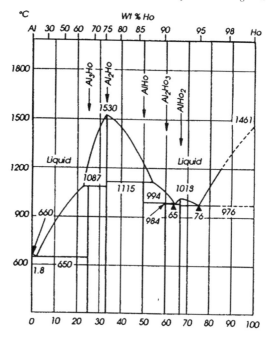

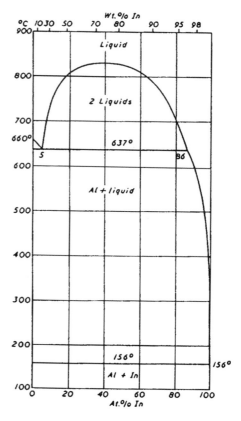

**Al–K**

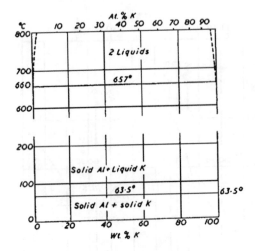

**Al–La**

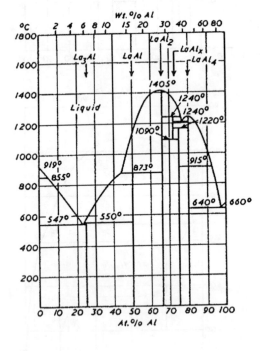

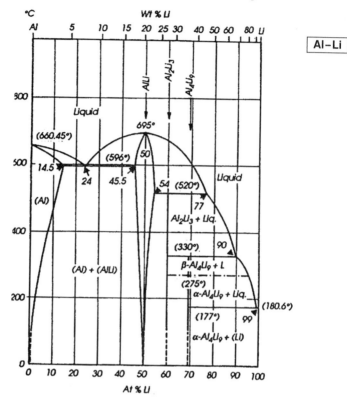

Al–Li

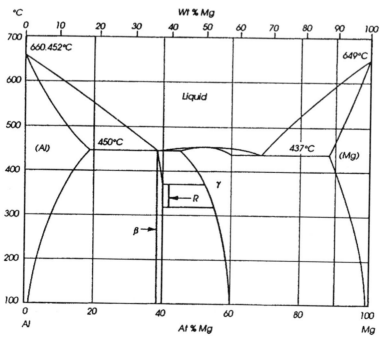

Al–Mg

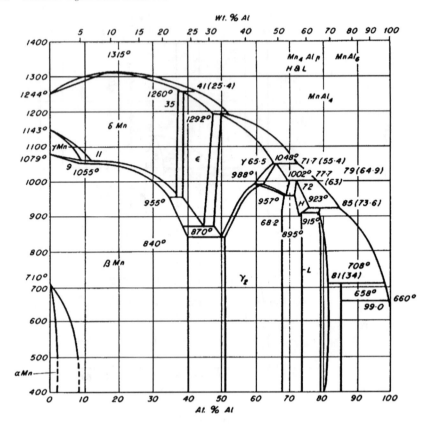

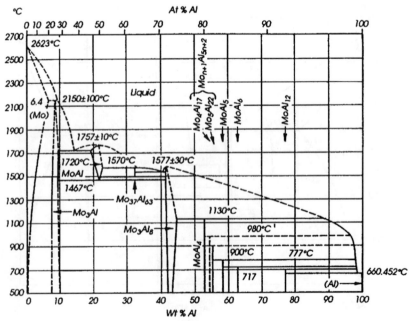

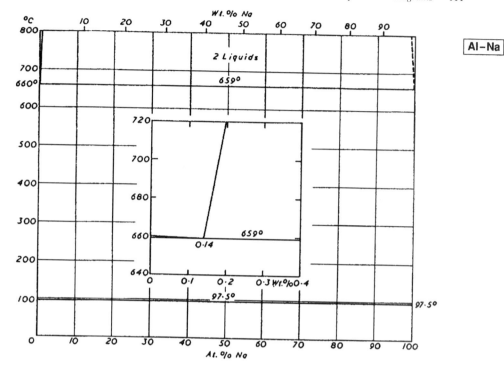

Al–Na

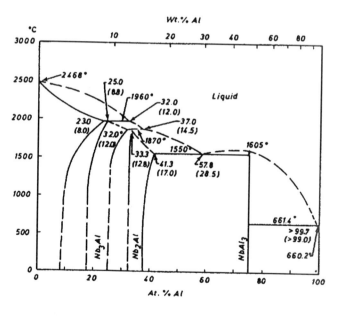

Al–Nb

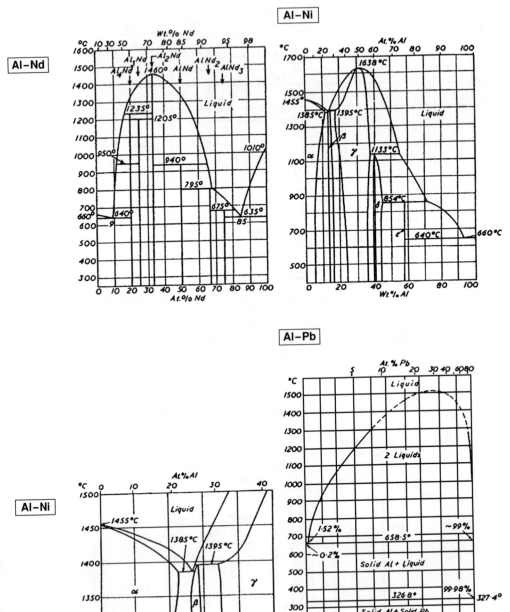

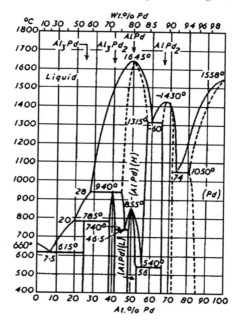

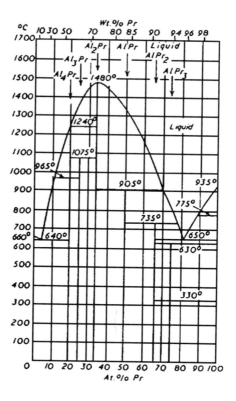

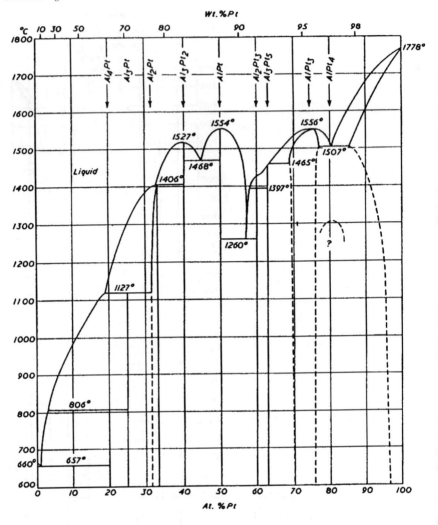

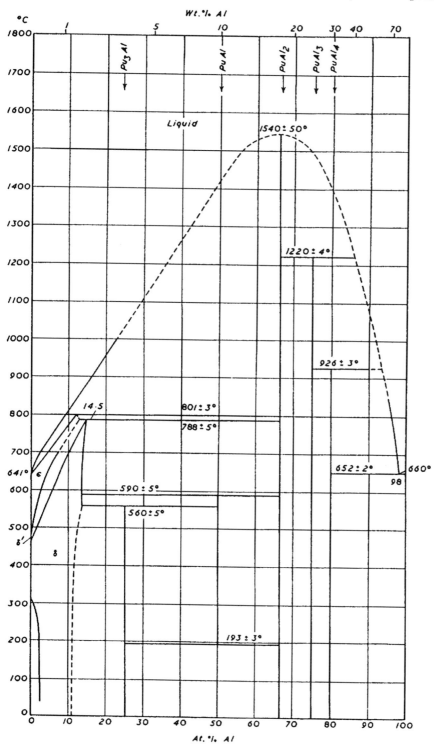

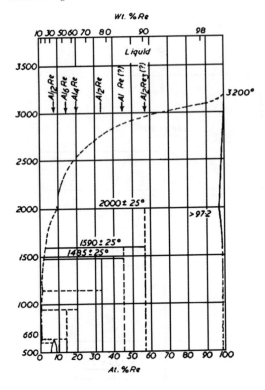

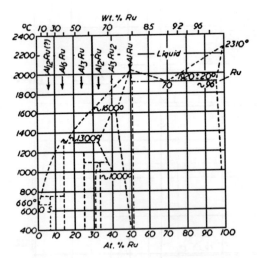

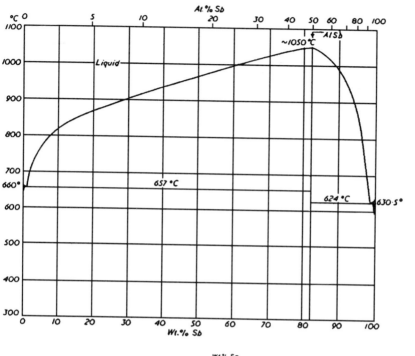

Al–Sb

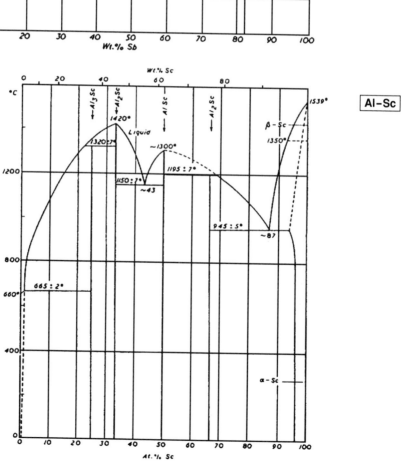

Al–Sc

**Al–Se**

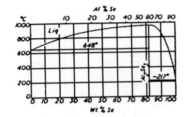

**Al–Si**

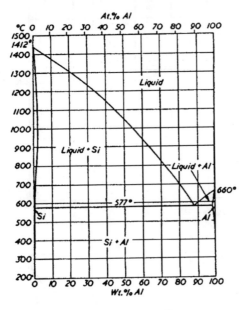

**Al–Sm**

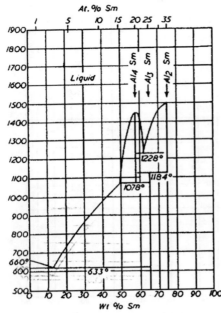

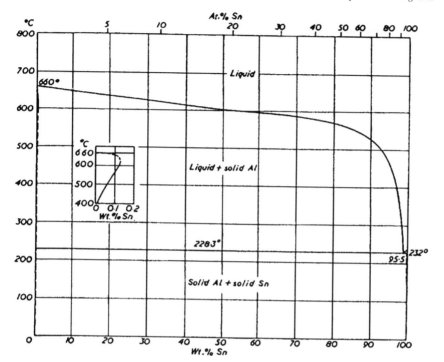

Al–Sn

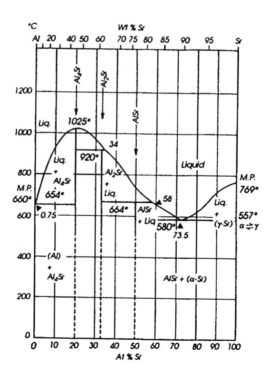

Al–Sr

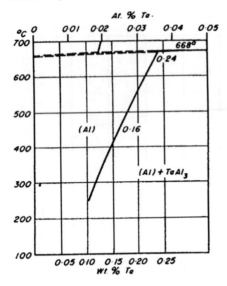

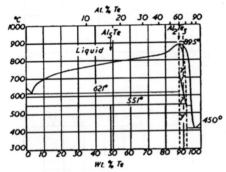

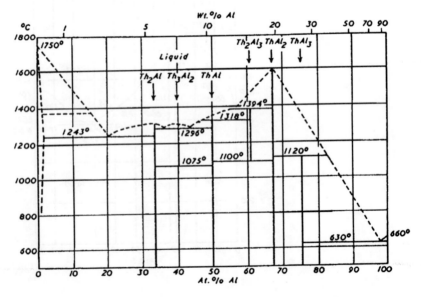

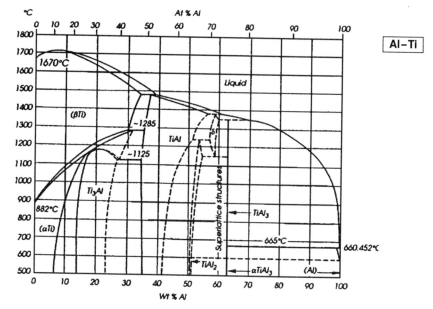

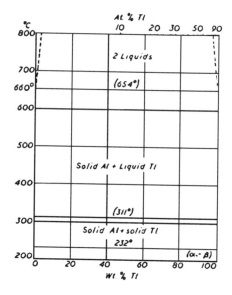

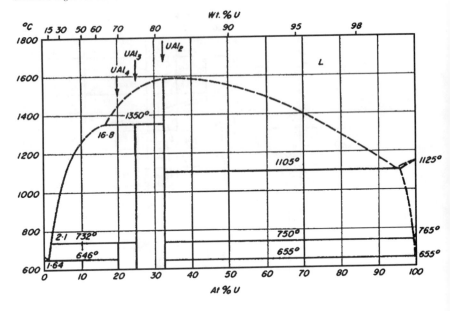

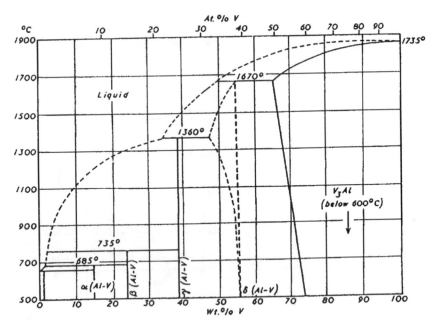

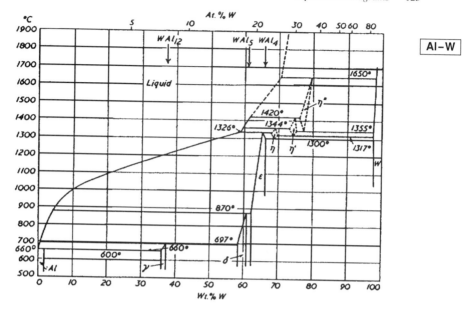

Al–W

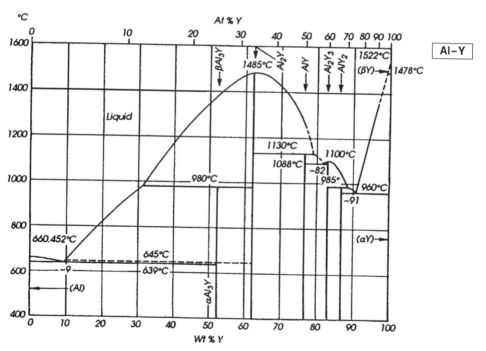

Al–Y

**Al–Yb**

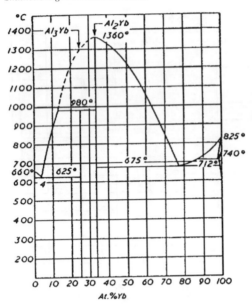

At.%Yb

**Al–Zn**

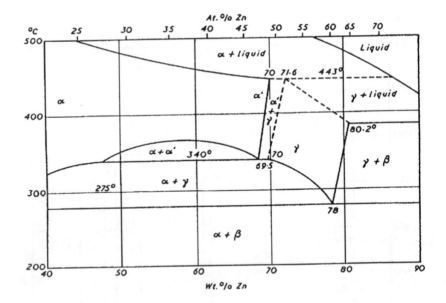

Wt.% Zn

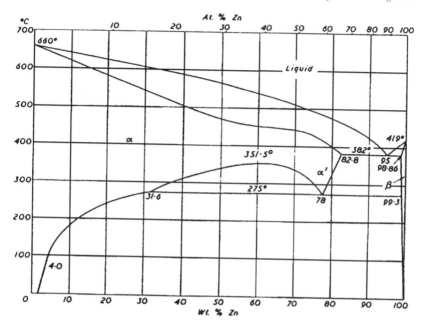

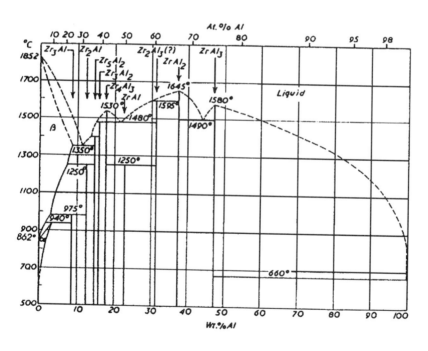

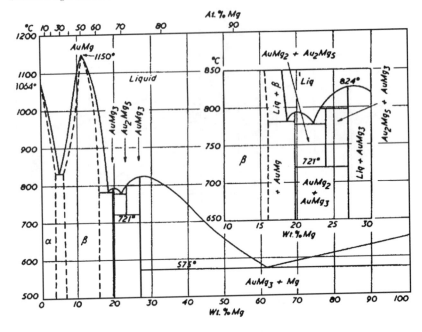

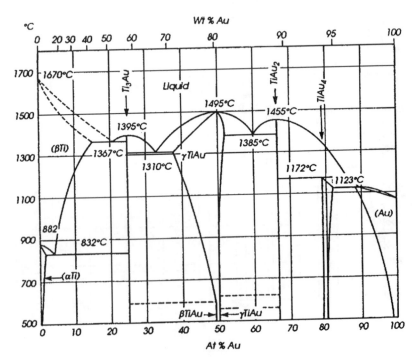

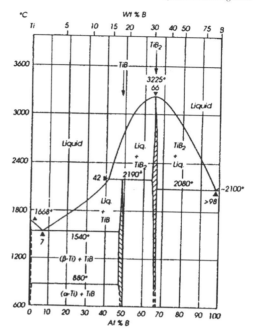

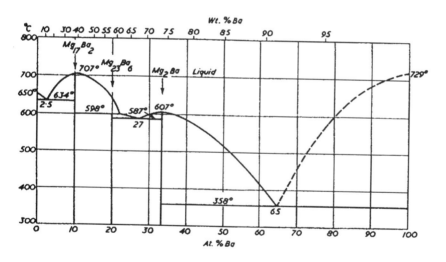

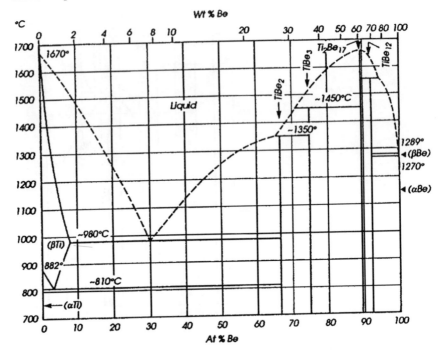

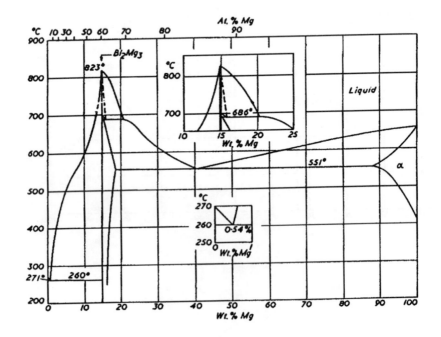

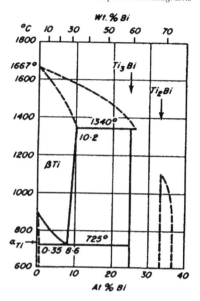

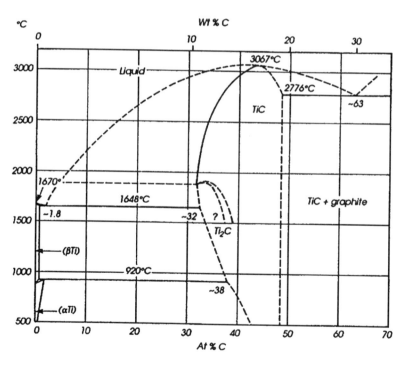

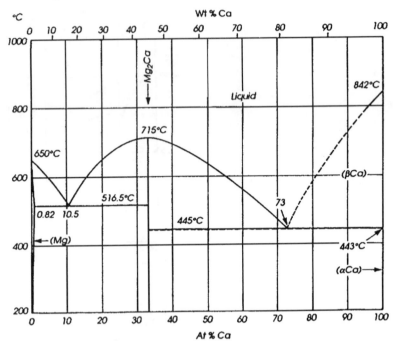

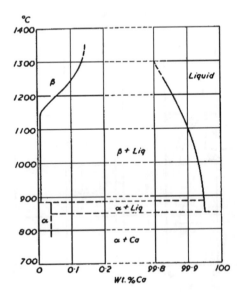

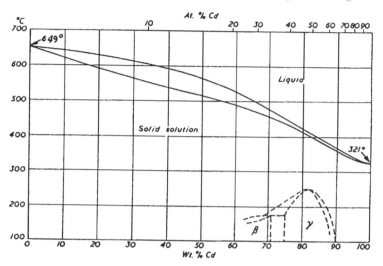

Cd–Mg

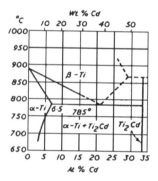

Cd–Ti

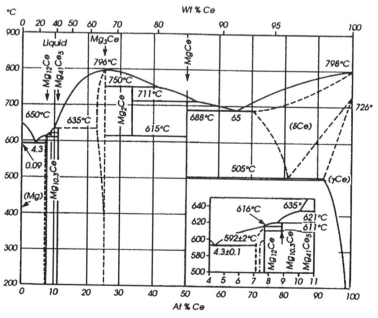

Ce–Mg

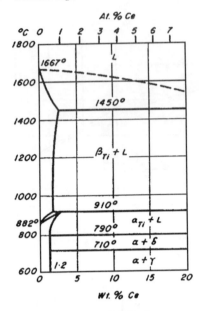

*Wt. % Ce*

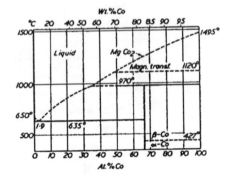

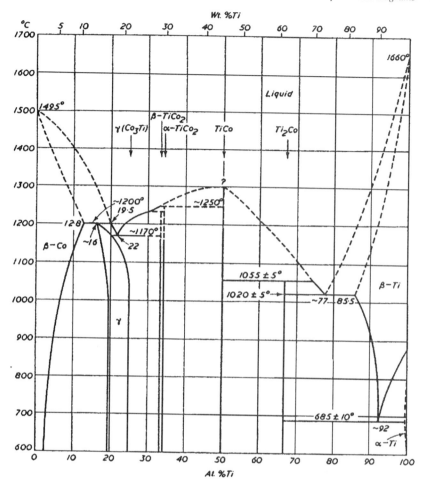

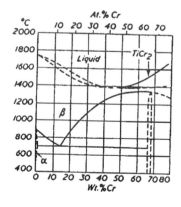

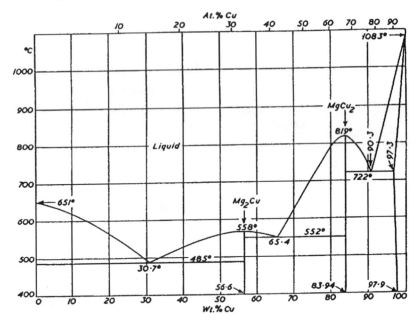

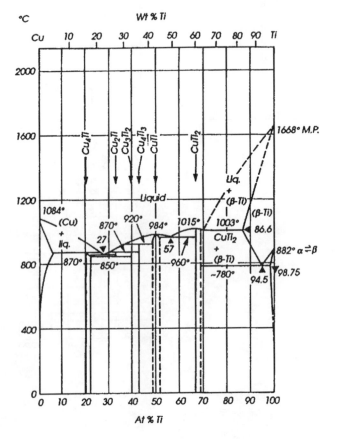

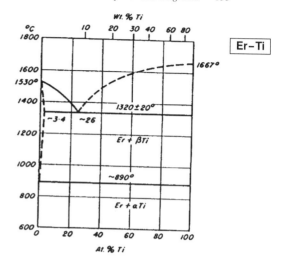

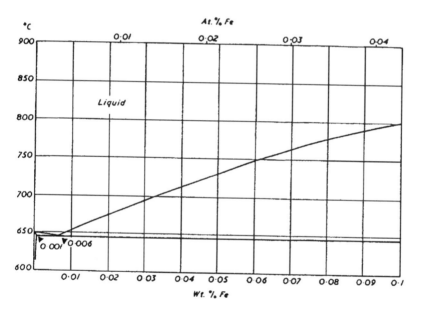

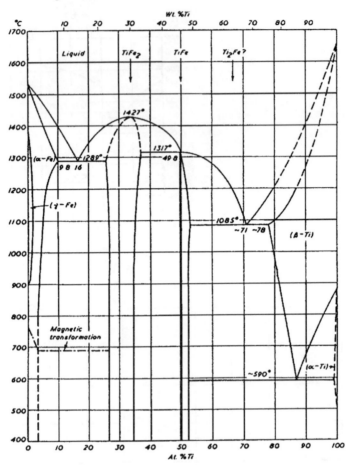

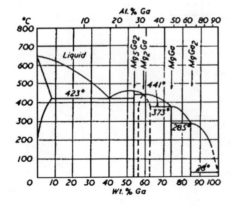

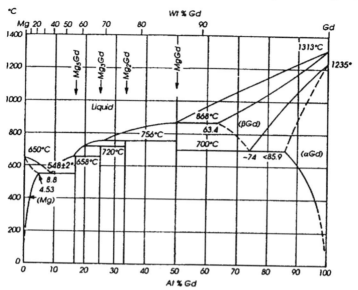

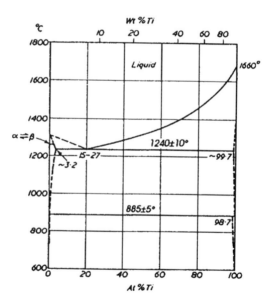

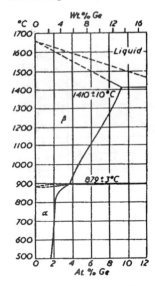

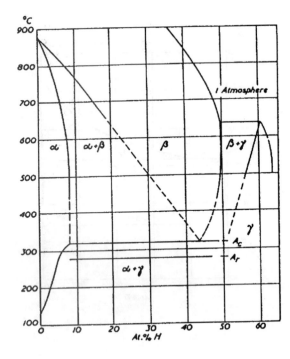

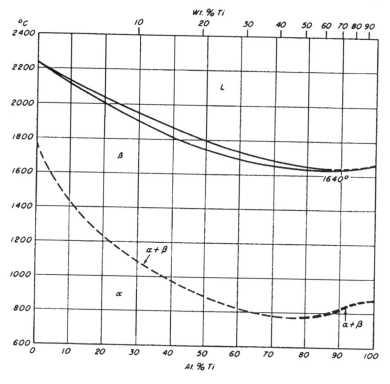

Hf–Ti

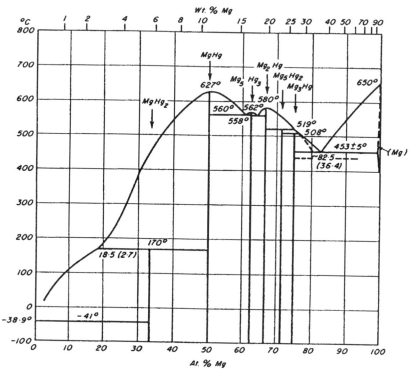

Hg–Mg

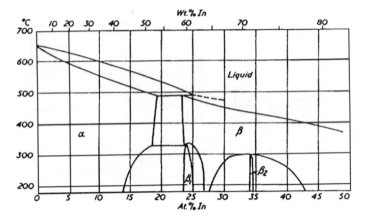

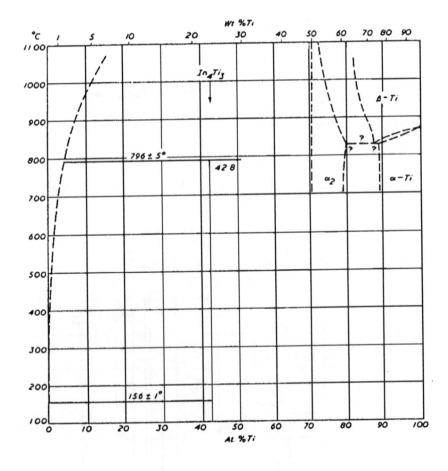

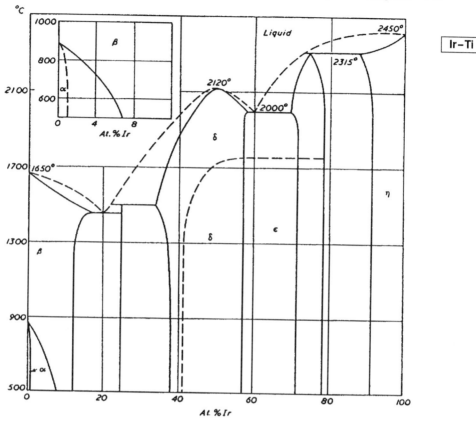

Ir–Ti

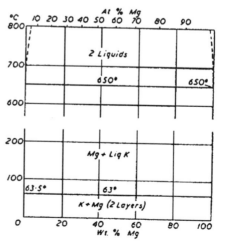

K–Mg

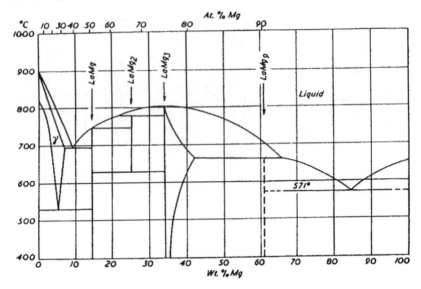

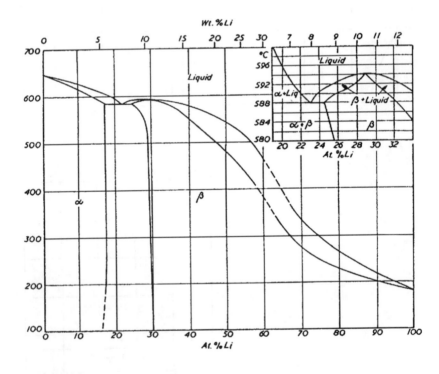

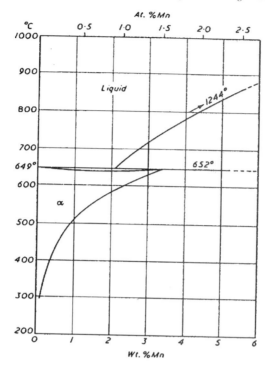

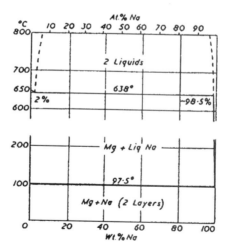

**Mg–Ni**

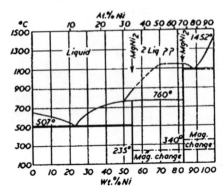

**Mg–Pb**

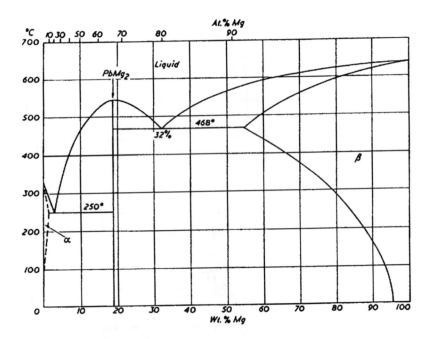

**Mg–Pr**

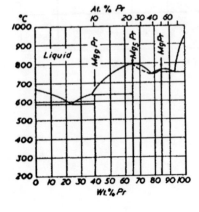

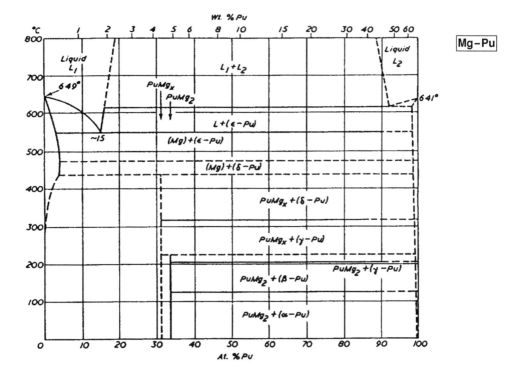

Mg–Pu

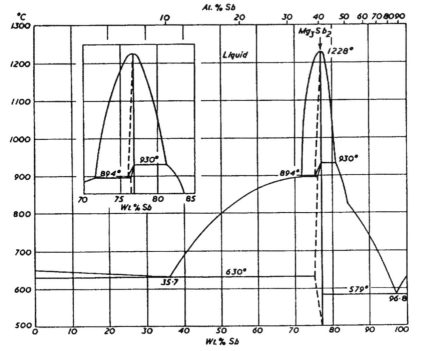

Mg–Sb

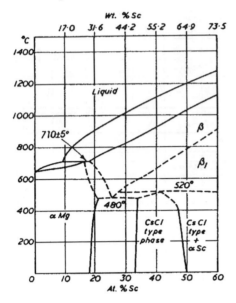

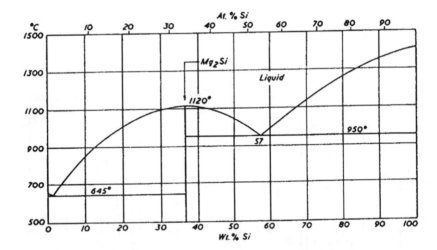

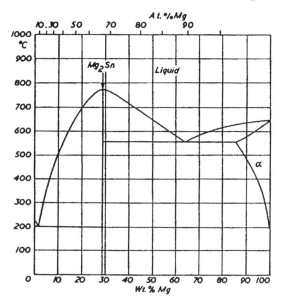

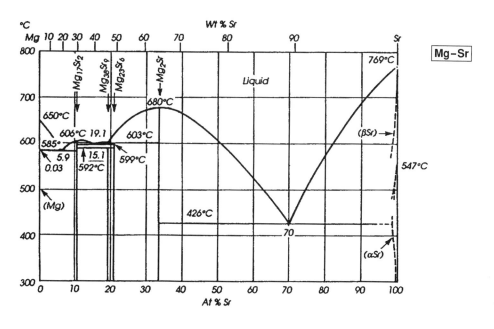

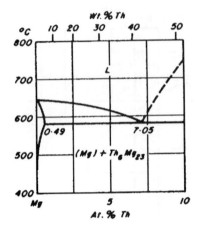

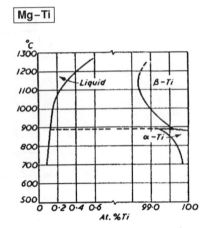

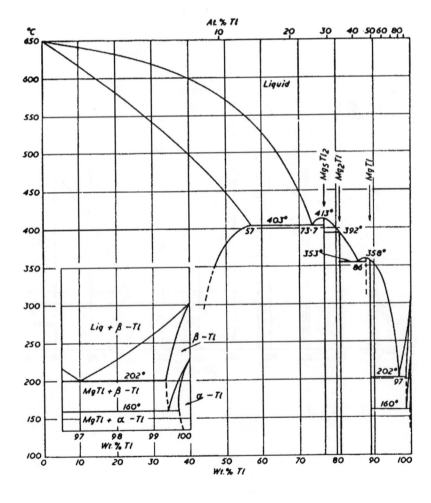

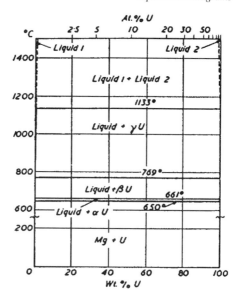

Mg–U

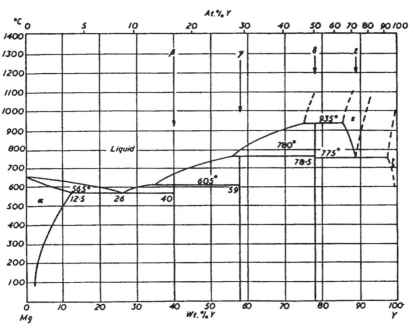

Mg–Y

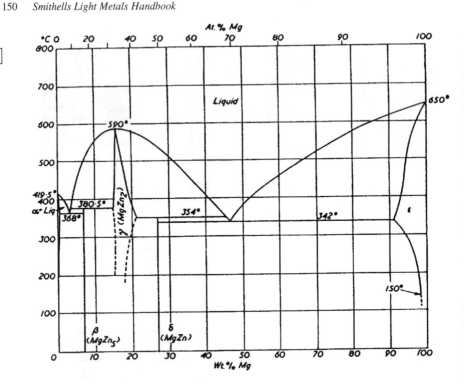

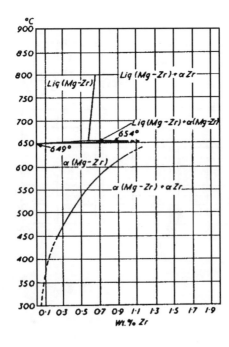

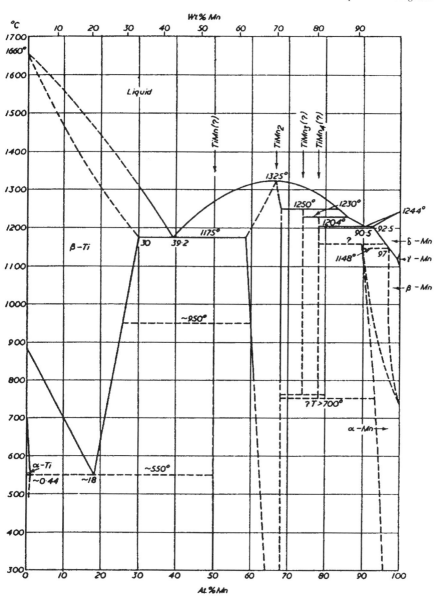

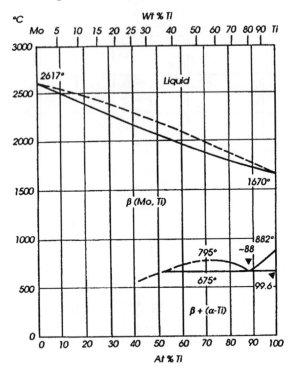

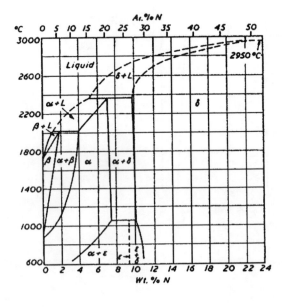

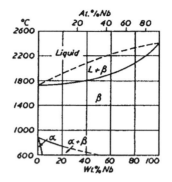

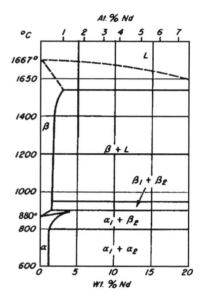

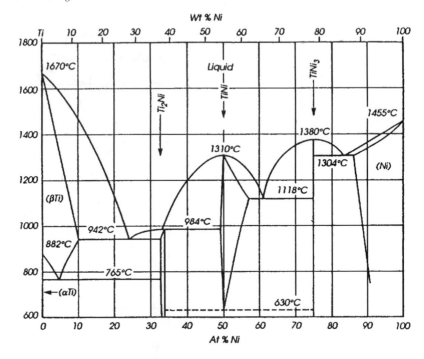

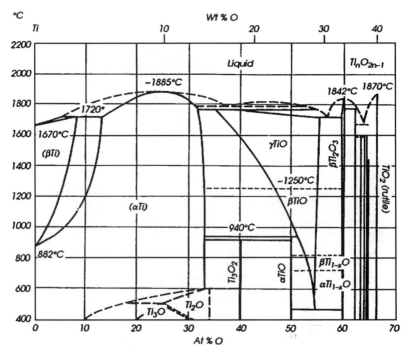

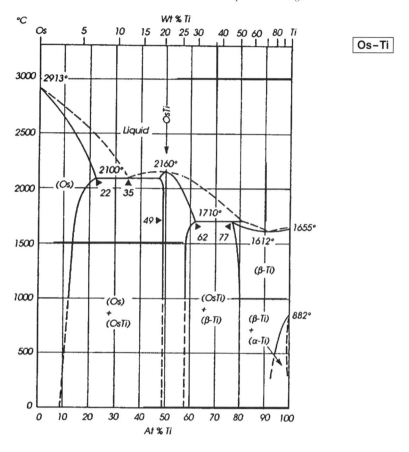

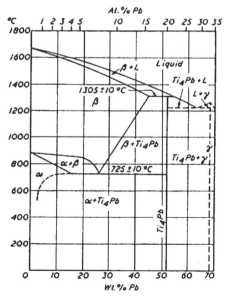

Pd–Ti

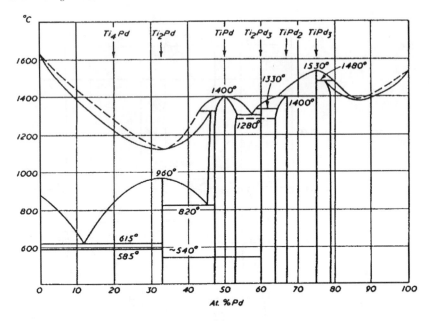

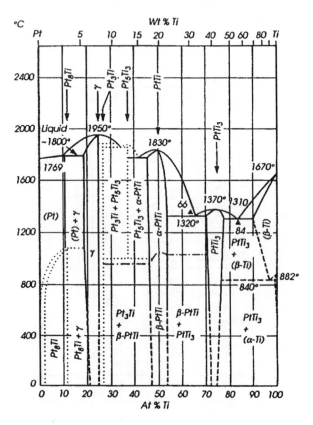

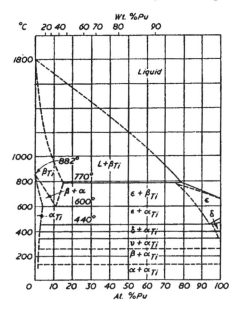

**Pu–Ti**

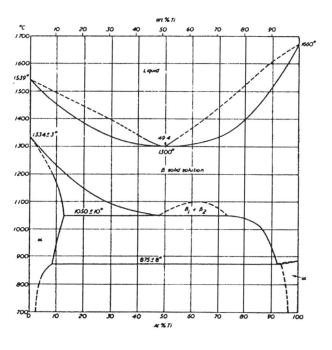

**Sc–Ti**

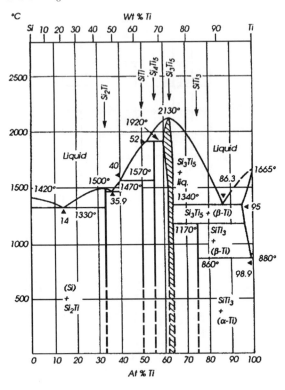

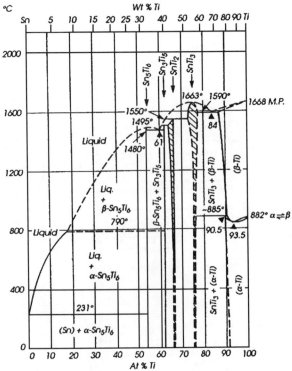

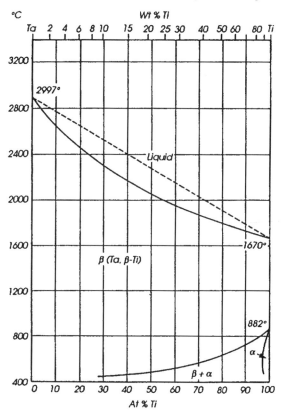

Ta–Ti

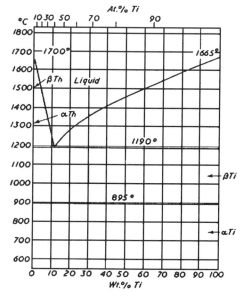

Th–Ti

Ti–U

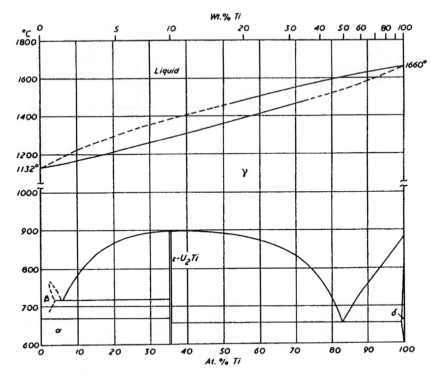

Ti–V

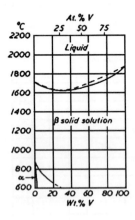

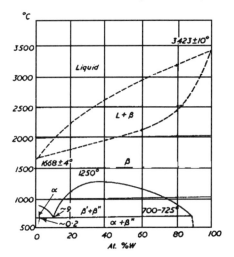

Ti–W

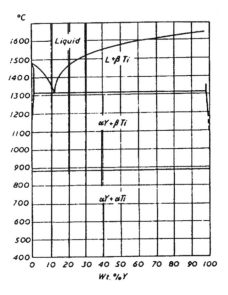

Ti–Y

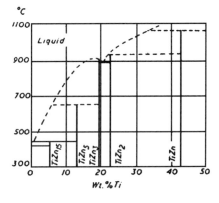

Ti–Zn

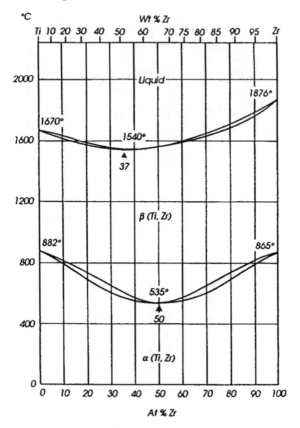

# 6 Metallography of light alloys

Metallography can be defined as the study of the structure of materials and alloys by the examination of specially prepared surfaces. Its original scope was limited by the resolution and depth of field in focus by the imaging of light reflected from the metallic surface. These limitations have been overcome by both transmission and scanning electron microscopy (TEM, STEM and SEM). The analysis of X-rays generated by the interaction of electron beams with atoms at or near the surface, by wavelength or energy dispersive detectors (WDX, EDX), has added quantitative determination of local composition, e.g. of intermetallic compounds, to the deductions from the well-developed etching techniques. Surface features can also be studied by collecting and analysing electrons diffracted from the surface. A diffraction pattern of the surface can be used to determine its crystallographic structure (low-energy electron diffraction or LEED). These electrons can also be imaged as in a conventional electron microscope (Low-energy electron microscopy or LEEM). This technique is especially useful for studying dynamic surface phenomena such as those occurring in catalysis. X-rays photoelectron microscopy (XPS or ESCA) now enables the metallographer to analyse the atoms in the outermost surface layer to a depth of a few atoms (0.3–5.0 nm) and provides information about the chemical environment of the atom. Auger spectroscopy uses a low-energy electron beam instead of X-rays to excite atoms, and analysis of the Auger electrons produced provides similar information about the atoms from which the Auger electron is ejected.

Nevertheless, the conventional optical techniques still have a significant role to play and their interpretation is extended and reinforced by the results of the electronic techniques.

## 6.1 Metallographic methods for aluminium alloys

PREPARATION

Aluminium and its alloys are soft and easily scratched or distorted during preparation. For cutting specimens, sharp saw-blades should be used with light pressure to avoid local overheating. Specimens may be ground on emery papers by the usual methods, but the papers should preferably have been already well used, and lubricated or coated with a paraffin oil ('white spirit' is suitable), paraffin wax or a solution of paraffin wax in paraffin oil. Silicon carbide papers (down to 600-grit) which can be well washed with water are preferred for harder alloys, the essential point being to avoid the embedding of abrasive particles in the metal. For pure soft aluminium, a high viscosity paraffin is needed to avoid this. Polishing is carried out in two stages: initial polishing with fine $\alpha$-alumina, proprietary metal polish, or diamond, and final polishing with $\gamma$-alumina or fine magnesia, using a slowly rotating wheel (not above 150 rev. min$^{-1}$). It is essential to use properly graded or levigated abrasives and it is preferable to use distilled water only; it is an advantage to boil new polishing cloths in water for some hours in order to soften the fibres. Many aluminium alloys contain hard particles of various intermetallic compounds, and polishing times should in general be as short as possible owing to the danger of producing excessive relief. Relief may be minimized by experience and skill in polishing; blanket felt may with advantage be substituted for velveteen or selvyt cloth as a polishing pad, while the use of parachute silk on a cork pad is also useful for avoiding relief in the initial stages of the process, but a better general alternative is to use diamond polishing, followed by a very brief final polishing with magnesia.

Many aluminium alloys contain the reactive compound $Mg_2Si$. If this constituent is suspected, white spirit should be substituted for water during all but the initial stages of wet polishing, to avoid loss of the reactive particles by corrosion.

**Table 6.1**   MICROCONSTITUENTS WHICH MAY BE ENCOUNTERED IN ALUMINIUM ALLOYS

| Microconstituent | Appearance in unetched polished sections |
|---|---|
| $Al_3Mg_2$ | Faint, white. Difficult to distinguish from the matrix. |
| $Mg_2Si$ | Slate grey to blue. Readily tarnishes on exposure to air and may show irridescent colour effects. Often brown if poorly prepared. Forms Chinese script eutectic. |
| $CaSi_2$ | Grey. Easily tarnished |
| $CuAl_2$ | Whitish, with pink tinge. A little in relief; usually rounded |
| $NiAl_3$ | Light grey, with a purplish pink tinge |
| $Co_2Al_9$ | Light grey |
| $FeAl_3^{(1)}$ | Lavender to purplish grey; parallel-sided blades with longitudinal markings |
| $MnAl_6$ | Flat grey. The other constituents of binary aluminium-manganese alloys ($MnAl_4$, $MnAl_3$ and '$\delta$') are also grey and appear progressively darker. May form hollow parallelograms |
| $CrAl_7$ | Whitish grey; polygonal. Rarely attacked by etches |
| Silicon | Slate grey. Hard, and in relief. Often primary with polygonal shape – use etch to outline |
| $\alpha(AlMnSi)^{(2)}$ | Light grey, darker and more buff than $MnAl_6$ |
| $\beta(AlMnSi)^{(2)}$ | Darker than $\alpha(AlMnSi)$, with a more bluish grey tint. Usually occurs in long needles |
| $Al_2CuMg$ | Like $CuAl_2$ but with bluish tinge |
| $Al_6Mg_4Cu$ | Flat, faint and similar to matrix |
| $(AlCuMn)^{(3)}$ | Grey |
| $\alpha(AlFeSi)^{(4)}$ | Purplish grey. Often occurs in Chinese-script formation. Isomorphous with $\alpha(AlMnSi)$ |
| $\beta(AlFeSi)^{(4)}$ | Light grey. Usually has a needle-like formation |
| $(AlCuFe)^{(5)}$ | Grey $\alpha$ phase lighter than $\beta$ phase (*see* Note 5) |
| $(AlFeMn)^{(6)}$ | Flat grey, like $MnAl_6$ |
| $(AlCuNi)$ | Purplish grey |
| $(AlFeSiMg)^{(7)}$ | Pearly grey |
| $FeNiAl_9$ | Very similar to and difficult to distinguish from $NiAl_3$ |
| $(AlCuFeMn)$ | Light grey |
| $Ni_4Mn_{11}Al_{60}$ | Purplish grey |
| $MgZn_2$ | Faint white; no relief |

In Table 10.1 constituents are designated by symbols denoting the compositions upon which they appear to be based, or by the elements, in parentheses, of which they are composed. The latter nomenclature is adopted where the composition is unknown, not fully established, or markedly variable. The superscript numbers in column 1 refer to the following notes:

(1) On very slow cooling under some conditions, $FeAl_3$ decomposes into $Fe_2Al_7$ and $Fe_2Al_5$. The former is micrographically indistinguishable from $FeAl_3$. The simpler formula is retained for consistency with most of the original literature.
(2) $\alpha(AlMnSi)$ is present in all slowly solidified aluminium-manganese-silicon alloys containing more than 0.3% of manganese and 0.2% of silicon, while $\beta(AlMnSi)$, a different ternary compound, occurs above approximately 3% of manganese for alloys containing more than approximately 1.5% of silicon. $\alpha(AlMnSi)$ has a variable composition in the region of 30% of manganese and 10–15% of silicon. The composition of $\beta(AlMnSi)$ is around 35% of manganese and 5–10% of silicon.
(3) $(AlCuMn)$ is a ternary compound with a relatively large range of homogeneity based on the composition $Cu_2Mn_3Al_{20}$.
(4) $\alpha(AlFeSi)$ may contain approximately 30% of iron and 8% of silicon, while $\beta(AlFeSi)$ may contain approximately 27% of iron and 15% of silicon. Both constituents may occur at low percentages of iron and silicon.
(5) The composition of this phase is uncertain. Two ternary phases exist. $\alpha(AlCuFe)$ resembles $FeAl_3$; $\beta(AlCuFe)$ forms long needles.
(6) The phase denoted as $(AlFeMn)$ is a solid solution of iron in $MnAl_6$.
(7) This constituent is only likely to be observed at high silicon contents.

It should be noted that some aluminium alloys are liable to undergo precipitation reactions at the temperatures used to cure thermosetting mounting resins; this applies particularly to aluminium-magnesium alloys, in which grain boundary precipitates may be induced.

ETCHING

The range of aluminium alloys now in use contains many complex alloy systems. A relatively large number of etching reagents have therefore been developed, and only those whose use has become more or less standard practice are given in Table 6.2. Many etches are designed to render the distinction between the many possible microconstituents easier, and the type of etching often depends on the magnification to be used. The identification of constituents, which is best accomplished by using cast specimens where possible, depends to a large extent on distinguishing between the

**Table 6.2**  ETCHING REAGENTS FOR ALUMINIUM AND ITS ALLOYS

| No. | Reagent | | Remarks |
|---|---|---|---|
| *1* | Hydrofluoric acid (40%) | 0.5 ml | 15 s immersion is recommended. Particles of all common micro-constituents are outlined. Colour indications: |
| | Hydrochloric acid (1.19) | 1.5 ml | |
| | Nitric acid (1.40) | 2.5 ml | $Mg_2Si$ and $CaSi_2$:  blue to brown |
| | Water | 95.5 ml | $\alpha(AlFeSi)$ and (AlFeMn):  darkened |
| | | | $\beta(AlCuFe)$:  light brown |
| | (Keller's etch)[†] | | $MgZn_2$, $NiAl_3$, (AlCuFeMn), $Al_2Cu\,Mg$ and  brown to black |
| | | | $Al_6CuMg$: |
| | | | $\alpha(AlCuFe)$ and (AlCuMn):  blackened |
| | | | $Al_3Mg_2$:  heavily outlined and pitted |
| | | | The colours of other constituents are little altered. Not good for high Si alloys |
| *2* | Hydrofluoric acid (40%) | 0.5 ml | 15 s swabbing is recommended. This reagent removes surface flowed layers, and reveals small particles of constituents, which are usually fairly heavily outlined. There is little grain contrast in the matrix. Colour indications: |
| | Water | 99.5 ml | |
| | | | $Mg_2Si$ and $CaSi_2$:  blue |
| | | | $FeAl_3$ and $MnAl_6$:  slightly darkened |
| | | | $NiAl_3$:  brown (irregular) |
| | | | $\alpha(AlFeSi)$:  dull brown |
| | | | (AlCrFe):  light brown |
| | | | $Co_2Al_9$:  dark brown |
| | | | (AlFeMn):  brownish tinge |
| | | | $\alpha(AlCuFe)$, (AlCuMg) and (AlCuMn):  blackened |
| | | | $\alpha(AlMnSi)$, $\beta(AlMnSi)$ and (AlCuFeMn) may appear light brown to black |
| | | | $\beta(AlFeSi)$ is coloured red brown to black |
| | | | The remaining possible constituents are little affected |
| *3* | Sulphuric acid | | 30 s immersion at 70 °C; the specimen is quenched in cold water. |
| | (1.84) | 20 ml | Colour indications: |
| | Water | 80 ml | $Mg_2Si$, $Al_3Mg_2$ and $FeAl_3$:  violently attacked, blackened and may be dissolved out |
| | | | $CaSi_2$:  blue |
| | | | $\alpha(AlMnSi)$ and $\beta(AlMnSi)$:  rough and attacked |
| | | | $NiAl_3$ and (AlCuNi):  slightly darkened |
| | | | $\beta(AlFeSi)$:  slightly darkened and pitted |
| | | | $\alpha(AlFeSi)$, (AlCuMg) and (AlCuFeMn): outlined and blackened |
| | | | Other constituents are not markedly affected |
| *4* | Nitric acid (1.40) | 25 ml | Specimens are immersed for 40 s at 70 °C and quenched in cold water. |
| | Water | 75 ml | Most constituents (not $MnAl_6$) are outlined. Colour indications: |
| | | | $\beta(AlCuFe)$ is slightly darkened |
| | | | $Al_3Mg_2$ and AlMnSi: attacked and darkened slightly |
| | | | $Mg_2Si$, $CuAl_2$, (AlCuNi) and (AlCuMg) are coloured brown to black |
| *5* | Sodium hydroxide | 1 g | Specimens are etched by swabbing for 10 s. All usual constituents are heavily outlined, except for $Al_3Mg_2$ (which may be lightly outlined) and (AlCrFe) which is both unattacked and uncoloured. Colour indications: |
| | Water | 99 ml | |
| | | | $FeAl_3$ and $NiAl_3$:  slightly darkened |
| | | | (AlCuMg):  light brown |
| | | | $\alpha(AlFeSi)$:  dull brown* |
| | | | $\alpha(AlMnSi)$:  rough and attacked; slightly darkened* |
| | | | $MnAl_6$ and (AlFeMn):  coloured brown to blue (uneven attack) |
| | | | $MnAl_4$:  tends to be darkened |
| | | | The colours of other constituents are only slightly altered |

*continued overleaf*

**Table 6.2**   *(continued)*

| No. | Reagent | | Remarks |
|---|---|---|---|
| 6 | Sodium hydroxide | 10 g | Specimens immersed for 5 s at 70 °C, and quenched in cold water. Colour indications: |
| | Water | 90 ml | $\beta$(AlFeSi): slightly darkened |
| | | | $Mn_{11} Ni_4Al_{60}$: light brown |
| | | | $\beta$(AlCuFe): light brown and pitted |
| | | | $CuAl_2$: light to dark brown |
| | | | $FeAl_3$: dark brown |
| | | | ($FeAl_3$ is more rapidly attacked in the presence of $CuAl_2$ than when alone) |
| | | | $MnAl_6$, $NiAl_3$, (AlFeMn), $CrAl_7$ and AlCrFe: blue to brown |
| | | | $\alpha$(AlFeSi), $\alpha$(AlCuFe), $CaSi_2$ and (AlCuMn): blackened |
| 7 | Sodium hydroxide | 3%–5% | Useful for sensitive etching where reproducibility is essential. In general, the effects are similar to those of Reagent 5, but the tendency towards colour variations for a given constituent is diminished. Particularly useful for distinguishing $FeNiAl_9$ (dark blue) from $NiAl_3$ brown). Potassium salts can be used. |
| | Sodium carbonate (in water) | 3%–5% | |
| 8 | Nitric acid | 20 ml | A reliable reagent for grain boundary etching, especially if the alternate polish and etch technique is adopted. The colours of particles are somewhat accentuated |
| | Hydrofluoric acid | 20 ml | |
| | Glycerol | 60 ml | |
| 9 | Nitric acid, 1% to 10% by vol. in alcohol | | Recommended for aluminium-magnesium alloys. $Al_3Mg_2$ is coloured brown. 5–20% chromium trioxide can be used |
| 10 | Picric acid | 4 g | Etching for 10 min darkens $CuAl_2$, leaving other constituents unaffected. Like reagent 4 |
| | Water | 96 ml | |
| 11 | Orthophosphoric acid | 9 ml | The reagent is used cold. Recommended for aluminium-magnesium alloys in which it darkens any grain boundaries containing thin $\beta$-precipitates. Specimen is immersed for a long period (up to 30 min). $Mg_2Si$ is coloured black, $Al_3Mg_2$ a light grey, and the ternary (AlMnFe) phase a dark grey |
| | Water | 91 ml | |
| 12 | Nitric acid | | 10 s immersion colours $Al_6CuMg_4$ greenish brown and distinguishes it from $Al_2CuMg$, which is slightly outlined but not otherwise affected |
| 13 | Nitric acid (density 1.2) | 20 ml | 20 ml of reagent are mixed with 80 ml alcohol. Specimens are immersed, and well washed with alcohol after etching. Brilliant and characteristic colours are developed on particles of intermetallic compounds. The effects depend on the duration of etching, and for differentiation purposes standardisation against known specimens is advised |
| | Water | 20 ml | |
| | Ammonium molybdate, $(NH_4)_6Mo_7O_{24}$, $4H_2O$ | 3 g | |
| 14 | Sodium hydroxide (various strengths, with 1 ml of zinc chloride per 100 ml of solution) | | Generally useful for revealing the grain structure of commercial aluminium alloy sheet[67] |
| 15 | Hydrochloric acid (37%) | 15.3 ml | Recommended (30 s immersion at room temperature) for testing the diffusion of copper through claddings of aluminium, aluminium-manganese-silicon, or aluminium-manganese on aluminium-copper-magnesium sheet. Zinc contents up to 2% in the clad material do not influence the result[68] |
| | Hydrofluoric acid (38%) | 7.7 ml | |
| | Water | 77.0 ml | |
| 16 | Ammonium oxalate | 1 g | Develops grain boundaries in aluminium-magnesium-silicon alloys. Specimens are etched for 5 min at 80 °C in a solution freshly prepared for each experiment |
| | Ammonium hydroxide, 15% in water | 100 ml | |

\*These are isomorphous and the colour depends on the proportion of Mn and Fe.

†Sodium fluoride can be used in place of HF in mixed acid etches.

colours of particles, so that the illumination should be as near as possible to daylight quality. It is recommended that a set of specially prepared standard specimens, containing various known metallographic constituents, be used for comparison.

It is very easy to obtain anomalous etching effects, such as ranges of colour in certain types of particles, and carefully standardized procedure is necessary. It should be remembered that the form and colour of the microconstituents may vary according to the degree of dispersion brought about by mechanical treatments, and also that the etching characteristics of a constituent may vary according to the nature of the other constituents present in the same section.

Some etching reagents for aluminium require the use of a high temperature; in such cases the specimen should be preheated to this temperature by immersion in hot water before etching. For washing purposes, a liberal stream of running water is advisable.

*Electrolytic etching for aluminium alloys.* In addition to the reagents given for aluminium in Table 6.2 the following solutions have been found useful for a restricted range of aluminium-rich alloys:

1. The following solution has been used for grain orientation studies:
   | | |
   |---|---|
   | Orthophosphoric acid (density 1.65) | 53 ml |
   | Distilled water | 26 ml |
   | Diethylene glycol monoethyl ether | 20 ml |
   | Hydrofluoric acid (48%) | 1 ml |

   The specimen should be at room temperature and electrolysis is carried out at 40 V and less than 0.1 A dm$^{-2}$. An etching time of 1.5–2 min is sufficient for producing grain contrast in polarized light after electropolishing.

2. The solution below is also used for the same purpose and is more reliable for some alloys:
   | | |
   |---|---|
   | Ethyl alcohol | 49 ml |
   | Water | 49 ml |
   | Hydrofluoric acid | 2 ml (quantity not critical) |

   The specimen is anodized in this solution at 30 V for 2 min at room temperature. A glass dish must be used. Not suitable for high-copper alloys.

3. For aluminium alloys containing up to 7% of magnesium:
   | | |
   |---|---|
   | Nitric acid (density 1.42) | 2 ml |
   | 40% hydrofluoric acid | 0.1 ml |
   | Water | 98 ml |

   Electrolysis is carried out at a current density of 0.3 A dm$^{-2}$ and a potential of 2 V. The specimen is placed 7.6 cm from a carbon cathode.

4. For cast duralumin:
   | | |
   |---|---|
   | Citric acid | 100 g |
   | Hydrochloric acid | 3 ml |
   | Ethyl alcohol | 20 ml |
   | Water | 977 ml |

   Electrolysis is carried out at 0.2 A dm$^{-2}$ and a potential of 12 V.

5. For commercial aluminium:
   | | |
   |---|---|
   | Hydrofluoric acid (40%) | 10 ml |
   | Glycerol | 55 ml |
   | Water | 35 ml |

   This reagent, used for 5 min at room temperature, with a current density of 1.5 A dm$^{-2}$ and a voltage of 7–8 V, is suitable for revealing the grain structure after electropolishing.[72]

6. For distinguishing between the phases present in aluminium-rich aluminium-copper-magnesium alloys, electrolytic etching in either ammonium molybdate solution or 0.880 ammonia has been recommended. In both cases, $Al_2CuMg$ is hardly affected, $CuAl_2$ is blackened, $Al_6Mg_4Cu$ is coloured brown, while $Mg_2Al_3$ is thrown into relief without change of colour.[73]

GRAIN-COLOURING ETCH

For many aluminium alloys containing copper, and especially for binary aluminium-copper alloys, it is found that Reagent No. 1 of Table 6.2 gives copper films on cubic faces which are subject to preferential attack and greater roughening of the surface. Subsequent etching with 1% caustic soda solution converts the copper into bronze-coloured cuprous oxide, and a brilliant and contrasting representation of the underlying surfaces is obtained. The technique is of use in orientation studies in so far as the films are dark and unbroken on (100) surfaces, but shrink on drying on other surfaces. In particular (111) faces have a bright yellow colour with a fine network on drying, which has no preferred orientation, while (110) faces develop lines (cracks in film) which are parallel to a cube edge.

## 6.2  Metallographic methods for magnesium alloys

PREPARATION

1. Magnesium is soft and readily forms mechanical twins and so deformed layers should be avoided.
2. Abrasives and polishing media tend to become embedded. Therefore use papers well-covered with paraffin making sure the deformed layer is removed.
3. Some phases in magnesium alloys are attacked by water. If these are present use paraffin or ethanol as lubricant.
4. Some very hard intermetallics can be present. Therefore keep polishing times short to avoid relief.

The recommended procedure is to grind carefully to 600-grit silicon carbide papers. Then polish with fine $\alpha$-alumina slurry or 4–6 μm diamond paste. This is followed by polishing on a fine cloth using light magnesia paste made with distilled water or a chemical attack polish of 1 g MgO, 20 ml ammon. tartrate soln. (10%) in 120 ml of distilled water. In reactive alloys, white spirit replaces distilled water and chemical attack methods avoided.

ETCHING

The general grain structure is revealed by examination under cross-polars. This will also detect mechanical twins formed during preparation. A selection of etching reagents suitable for magnesium and its alloys is given in Table 6.3. Of these, 4 and 1 are the most generally useful reagents for cast alloys, while 16 is a useful macro-etchant and, followed by 4, is invaluable for showing up the grain structure in wrought alloys.

**Table 6.3**  ETCHING REAGENTS FOR MAGNESIUM AND ITS ALLOYS

| No. | Etchant | | Remarks |
|---|---|---|---|
| 1 | Nitric acid | 1 ml | This reagent is recommended for general use, particularly with cast, |
| | Diethylene glycol | 75 ml | die-cast and aged alloys. Specimens are immersed for 10–15 s, |
| | Distilled water | 24 ml | and washed with hot distilled water. The appearance of common constituents following this treatment is outlined in Table 6.4. Mg-RE and Mg-Th alloys also |
| 2 | Nitric acid | 1 ml | Recommended for solution-heat-treated castings, and wrought |
| | Glacial acetic acid | 20 ml | alloys. Grain boundaries are revealed. The proportions are some- |
| | Water | 19 ml | what critical. Use 1–10 s |
| | Diethylene glycol | 60 ml | |
| 3 | Citric acid | 5 g | This reagent reveals grain boundaries, and should be applied by |
| | Water | 95 ml | swabbing. Polarized light is an alternative |
| 4 | Nitric acid, 2% in alcohol | | A generally useful reagent |
| 5 | Nitric acid, 8% in alcohol | | Etching time 4–6 s. Recommended for cast, extruded and rolled magnesium-manganese alloys |
| 6 | Nitric acid, 4% in alcohol | | Used for magnesium-rich alloys containing other phases, which are coloured light to dark brown |
| 7 | Nitric acid, 5% in water | | Etching time 1–3 s. Recommended for cast and forged alloys containing approximately 9% of aluminium |
| 8 | Oxalic acid 20 g l$^{-1}$ in water | | Etching time 6–10 s. Used also for extruded magnesium-manganese alloys |
| 9 | Acetic acid, 10% in water | | Etching time 3–4 s. Used for magnesium-aluminium alloys with 3% of aluminium |

**Table 6.3** *(continued)*

| No. | Etchant | Remarks |
|-----|---------|---------|
| *10* | Tartaric acid 20 g l$^{-1}$ of water | Etching time 6 s ⎫ These reagents are recommended for |
| *11* | Orthophosphoric acid, 13% in glycerol | Etching time 12 s ⎭ magnesium-aluminium alloys with 3 to 6% of aluminium |
| *12* | Tartaric acid 100 g l$^{-1}$ of water | Used for wrought alloys. $Mg_2Si$ is roughened and pitted. 10 s to 2 min for Mg-Mn-Al-Zn alloys. Grain contrast in cast alloys |
| *13* | Citric acid and nitric acid in glycerol | Used for magnesium-cerium and magnesium-zirconium alloys. The magnesium-rich matrix is darkened and the other phases left white |
| *14* | Orthophosphoric acid    0.7 ml<br>Picric acid    4 g<br>Ethyl alcohol    100 ml | Recommended for solution-heat-treated castings. The specimen is lightly swabbed, or immersed with agitation for 10–20 s. The magnesium-rich matrix is darkened, and other phases (except $Mg_2Sn$) are little affected. The maximum contrast between the matrix and $Mg_{17}Al_{12}$ is developed. The darkening of the matrix is due to the development of a film, which must not be harmed by careless drying |
| *15* | Picric acid saturated in 95% alcohol    10 ml<br>Glacial acetic acid    1 ml | A grain boundary etching reagent; especially for Dow metal (Al 3% Zn 1%, Mn 0.3%). Reveals cold work and twins |
| *16* | Picric acid, 5% in ethyl alcohol    50 ml<br>Glacial acetic acid    20 ml<br>Distilled water    20 ml | Useful for magnesium-aluminium-zinc alloys. On etching for 15 s an amorphous film is produced on the polished surface. When dry, the film cracks parallel to the trace of the basal plane in each grain. The reagent may be used to reveal changes of composition within grains, and other special purposes |
| *17* | Picric acid, 5% in ethyl alcohol    50 ml<br>Glacial acetic acid    16 ml<br>Distilled water    20 ml | As for Reagent 16, but suitable for a more restricted range of alloy composition |
| *18* | Picric acid, 5% in ethyl alcohol    100 ml<br>Glacial acetic acid    5 ml<br>Nitric acid (1.40)    3 ml | General reagent |
| *19* | Picric acid, 5% in ethyl alcohol    10 ml<br>Distilled water    10 ml | $Mg_2Si$ is coloured dark blue and manganese-bearing constituents are left unaffected |
| *20* | Hydrofluoric acid (40%)    10 ml<br>Distilled water    90 ml | Useful for magnesium-aluminium-zinc alloys. $Mg_{17}Al_{12}$ is darkened, and $Mg_3Al_2Zn_3$ is left unetched. If the specimen is now immersed in dilute picric acid solution (1 vol. of 5% picric acid in alcohol and 9 vol. of water) the matrix turns yellow, and the ternary compound remains white |
| *21* | Picric acid, 5% in ethyl alcohol    100 ml<br>Distilled water    10 ml<br>Glacial acetic acid    5 ml | Reveals grain-boundaries in both cast and wrought alloys. This reagent is useful for differentiating between grains of different orientations, and for revealing internally stressed regions |
| *22* | Nitric acid conc. | Recommended for pure metal only. Specimen is immersed in the cold acid. After 1 min a copious evolution of $NO_2$ occurs, and then almost ceases. At the end of the violent stage, the specimen is removed, washed and dried. Surfaces of very high reflectivity result, and grain boundaries are revealed |

*The appearance of constituents after etching.* The micrographic appearances of the commonly occurring microconstituents in cast alloys are as given in Table 6.4.

ELECTROLYTIC ETCHING OF MAGNESIUM ALLOYS

This has been recommended for forged alloys. The specimen is anodically treated in 10% aqueous sodium hydroxide containing 0.06 g l$^{-1}$ of copper. A copper cathode is used, and a current density of

**Table 6.4**  THE MICROGRAPHIC APPEARANCE OF CONSTITUENTS OF MAGNESIUM ALLOYS

| Microconstituent | Appearance in polished sections, etched with Reagent 1 (zirconium-free alloys) |
|---|---|
| $Mg_{17}Al_{12}^{(1)}$ | White, sharply outlined and brought into definite relief |
| $MgZn_2^{(2)}$ | Appearance very similar to that of $Mg_{17}Al_{12}$ |
| $Mg_3Al_2Zn_3^{(3)}$ | Appearance similar to those of $Mg_{17}Al_{12}$ and $MgZn_2$ |
| $Mg_2Si^{(4)}$ | Watery blue green; the phase usually has a characteristic Chinese-script formation, but may appear in massive particles. Relief less than for manganese |
| $Mg_2Sn^{(4)}$ | Tan to brown or dark blue, depending on duration of etching. Individual particles may differ in colour |
| Manganese$^{(5)}$ | Grey particles, usually rounded and in relief. Little affected by etching |
| $(MgMnAl)^{(5)}$ | Grey particles, angular in shape and in relief. Little affected by etching |

| Microconstituent | Appearance in polished sections etched with Reagent 4 (zirconium-bearing alloys) |
|---|---|
| Primary Zr (undissolved in molten alloy) | Hard, coarse, pinkish grey rounded particles, readily visible before etching |
| Zinc-rich particles$^{(6)}$ | Fine, dark particles, loosely clustered and comparatively inconspicuous before etching |
| $Mg_9Ce$ | Compound or divorced eutectic in grain boundaries. Appearance hardly changed by few per cent of zinc or silver |
| $Mg_5Th$ | Compound or divorced eutectic in grain boundaries (bluish). Appearance hardly changed by few per cent of zinc if Zn exceeds Th |
| Mg-Th-Zn | Brown acicular phase. Appears in Mg-Th-Zn-Zr alloys when Th $\geq$ Zn |
| $MgZn_2$ | Compound or divorced eutectic in grain boundaries. Absent from alloys containing RE or Th |

The superscript numbers in column 1 refer to the following notes:

(1) This is the $\gamma$-phase of the magnesium-aluminium system; it is also frequently called $Mg_4Al_3$ or $Mg_3Al_2$.
(2) Although the phase MgZn may be observed in equilibrium conditions, $MgZn_2$ is frequently encountered in cast alloys.
(3) This ternary compound occurs in alloys based on the ternary system magnesium-aluminium-zinc, and may be associated with $Mg_{17}Al_{12}$.
(4) Blue unetched.
(5) These constituents are best observed in the unetched condition.
(6) Alloys of zirconium with interfering elements such as Fe, Al, Si, N and H, separating as a Zr-rich precipitate in the liquid alloy. Co-precipitation of various impurities makes the particles of indefinite composition.

*Note*: The microstructure of all zirconium-bearing cast alloys with satisfactory dissolved zirconium content is characterised by Zr-rich coring in the centre of most grains. In the wrought alloys zirconium is precipitated from the cored areas during preheating or working, resulting in longitudinal striations of fine precipitate which become visible on etching.

$0.53\ A\,dm^{-2}$ is applied at 4 V. After etching, the specimen is successively washed with 5% sodium hydroxide, distilled water and alcohol, and is finally dried.

NON-METALLIC INCLUSIONS IN MAGNESIUM-BASE ALLOYS

The detection and identification of accidental flux and other inclusions in magnesium alloys involves the exposure of a prepared surface to controlled conditions of humidity, when corrosion occurs at the site of certain inclusions, others being comparatively unaffected. The corrosion product or the inclusion may then be examined by microchemical techniques.

The surface to be examined should be carefully machined and polished by standard procedures. The polishing time should be short, and alcohol or other solvent capable of dissolving flux must be avoided. As soon as possible the prepared specimens are placed in a humidity chamber, having been protected in transit by wrapping in paper. A suitable degree of humidity is provided by the air above a saturated solution of sodium thiosulphate. The presence of corrosive inclusions is indicated by the development of corrosion spots. At this stage the corroded area may be lightly ground away to expose the underlying structure for microexamination so that the micrographic features which are holding the flux become visible. With other specimens, or with the same specimens re-exposed to the humid conditions, identification of the inclusions may be proceeded with, as follows:

*1. Detection of chloride*

The corrosion product is scraped off, and dissolved on a microscope slip in 5% aqueous nitric acid. A 1% silver nitrate solution is then added, and a turbidity of silver chloride indicates the presence of the chloride ion. The solution of the corrosion product should preferably be heated before adding the silver nitrate to remove any sulphide ion, which also gives rise to turbidity. Alternatively, a 10% solution of chromium trioxide may be added directly to the corrosion spot, when chloride is indicated by an evolution of gas bubbles from the metal surface, and the development of a brown stain. This method is less specific than the silver nitrate method, and may give positive reactions in the presence of relatively large amounts of sulphates and nitrates.

### 2. Detection of calcium

Scrapings of corrosion product are dissolved in a small watch glass on a hot plate in 2 ml water and one drop of glacial acetic acid. To the hot solution a few drops of saturated ammonium oxalate solution are added. The presence of calcium is indicated by turbidity or precipitation. Spectroscopic identification of calcium in the solution is also possible.

### 3. Detection of boric acid in inclusions

Scrapings of corrosion product and metal are placed in a test tube with 1 ml of water. The inclusion dissolves, and complete solution of the sample is effected by adding a small portion of sulphuric acid (density 1.84) from 9 ml carefully measured and contained in a graduated cylinder. When solution is complete, the remainder of the acid is added and the mixture is well shaken; 0.5 ml of a 0.1% solution of quinalizarin in 93% (by wt.) of sulphuric acid is now added, mixed in, and allowed to stand for 5 min. A blue colour indicates the presence of boric acid. The colour in the absence of boric acid varies from bluish violet to red according to the dilution of the acid, which must thus be carefully controlled as described.

### 4. Detection of nitride

A drop of Nessler's solution applied directly to the metal surface in the presence of nitride, gives an orange brown precipitate, which may take about 1 min to develop. This test should be made on freshly prepared surfaces on which no water has been used, since decomposition of nitride to oxide occurs in damp air.

### 5. Detection of sulphide

The corrosion product is added to a few drops of water slightly acidified with nitric acid. A drop of the solution placed on a silver surface gives rise to a dark stain if sulphide was present in the corrosion product. Sulphur printing may also be applied.

### 6. Detection of iron

The corrosion product is dissolved in hydrochloric acid. A drop of nitric acid is added with several drops of distilled water. In the presence of iron, the addition of a crystal of ammonium thiocyanate develops a blood-red colouration.

In all the above tests, a simultaneous *blank* test should be carried out.

Iron-printing, analogous to sulphur-printing, can be applied using cleaned photographic paper impregnated with a freshly prepared solution of potassium ferricyanide and potassium ferrocyanide acidified with nitric or hydrochloric acid.

## 6.3   Metallographic methods for titanium alloys

PREPARATION

The preparation of titanium samples by ordinary methods of grinding is straightforward but needs care; final polishing is difficult. Specimens are easily scratched, and mechanical working of the surface during polishing causes twin-formation which may obscure other metallographic features. Other 'false' structures may be caused by the presence of local, randomly dispersed areas of cold work, which give a duplex appearance to homogeneous specimens. Electrolytic polishing of surfaces ground wet by ordinary methods to the 000 grade of emery paper is therefore recommended.

Mechanical polishing, if preferred, may be carried out with diamond preparation, with a final fine polish (if required) with alumina, both with a trace of hydrofluoric acid.

Examination for hydride is carried out in polarised light between crossed polaroids; the hydride then appears bright and anisotropic. This also reveals the grain structure of $\alpha$-titanium.

ETCHING

The presence of surface oxide films on titanium and its alloys necessitates the use of strongly acid etchants. Those given in Table 6.5 are useful.

**Table 6.5**   ETCHING REAGENTS FOR TITANIUM AND ITS ALLOYS

| No. | Etchant | | Conditions | Remarks |
|---|---|---|---|---|
| 1 | Hydrofluoric acid (40%) | 1–3 ml | 5–30 s | Mainly unalloyed titanium; reveals hydrides |
|  | Nitric acid (1.40) | 10 ml | | |
|  | Lactic acid | 30 ml | | |
| 2 | Hydrofluoric acid (40%) | 1 ml | 5–30 s | As Etchant 1 |
|  | Nitric acid (1.40) | 30 ml | | |
|  | Lactic acid | 30 ml | | |
| 3 | Hydrofluoric acid (40%) | 1–3 ml | 3–10 s | Most useful general etch |
|  | Nitric acid (1.40) | 2–6 ml | | |
|  | Water | to 100 ml | | |
|  | (Kroll's reagent) | | | |
| 4 | Hydrofluoric acid (40%) | 10 ml | 5–30 s | Chemical polish and g.b. etch |
|  | Nitric acid (1.40) | 10 ml | | |
|  | Lactic acid | 30 ml | | |
| 5 | Potassium hydroxide (40%) | 10 ml | 3–20 s | Useful for $\alpha/\beta$ alloys. $\alpha$ is attacked or stained. $\beta$ unattacked |
|  | Hydrogen peroxide (30%) | 5 ml | | |
|  | Water | 20 ml | | |
|  | (can be varied to suit alloy) | | | |
| 6 | Hydrofluoric acid (40%) | 20 ml | 5–15 s | General purpose, TiAlSn alloys |
|  | Nitric acid | 20 ml | | |
|  | Glycerol | 40 ml | | |
| 7 | Hydrofluoric acid | 1 ml | 3–20 s | TiAlSn alloys |
|  | Nitric acid (1.40) | 25 ml | | |
|  | Glycerol | 45 ml | | |
|  | Water | 20 ml | | |

# 7  Heat treatment of light alloys

## 7.1  Aluminium alloys

### 7.1.1  Annealing

For softening aluminium alloys that have been hardened by cold work:
Alloys 1080A, 1050, 1200, 5251, 5154A, 5454, 5083 – 360 °C for 20 min.
Alloys 3103, 3105 – 400–425 °C for 20 min.
Heat-treatable alloys that have not been heat treated – 360 °C ± 10 °C for 1 h and cool in air.
Alloys that have been heat treated – 400–425 °C for 1 h and cool at 15 °C/h to 300 °C.
For Al-Zn-Mg alloys of the 7000 series, after cooling in air, reheat to 225 °C for 2–4 h.

### 7.1.2  Stabilizing

To relieve internal stress. Normally heat to 250 °C followed by slow cooling is adequate.

### 7.1.3  Hardening

Conditions for solution treatment and ageing for both cast and wrought aluminium alloys are given in Tables 7.1 and 7.2. For the alloy designation system and compositions see Chapter 3. Temper designations are given in Table 7.3.

**Table 7.1**  HEAT TREATMENT DATA FOR ALUMINIUM CASTING ALLOYS

| Material designation and temper | Alloy type | Solution treatment | | | Precipitation treatment | |
|---|---|---|---|---|---|---|
| | | Temperature °C | Time[1] h | Quench[2] medium | Temperature °C | Time[3] h |
| **BS 1490** | | | | | | |
| LM 4–TF | Al Si5 Cu3 | 505–520 | 6–16 | Hot water | 150–170 | 6–18 |
| LM 9–TE | Al Si12 Mg | – | – | – | 150–170 | 16 |
| –TF | | 520–535 | 2–8 | Water | 150–170 | 16 |
| LM 10–TB | Al Mg10 | 425–435 | 8 | Oil at 160 °C max[4] | – | – |
| LM 13–TE | Al Si11 Mg Cu | – | – | – | 160–180 | 4–16[†] |
| –TF | | 515–525 | 8 | Hot water | 160–180 | 4–16 |
| –TF7 | | 515–525 | 8 | Hot water | 200–250 | 4–16 |
| LM 16–TB | Al Si5 Cu1 Mg | 520–530 | 12 | Hot water | – | – |
| –TF | | 520–530 | 12 | Hot water | 160–179 | 8–10 |
| LM 22–TB | Al Si6 Cu3 Mn | 515–530 | 6–9 | Hot water | – | – |

*continued overleaf*

**Table 7.1**   *(continued)*

| Material designation and temper | Alloy type | Solution treatment Temperature °C | Time[1] h | Quench[2] medium | Precipitation treatment Temperature °C | Time[3] h |
|---|---|---|---|---|---|---|
| LM 25–TB7 | Al Si7 Mg | 525–545 | 4–12 | Hot water | 250 | 2–4 |
| –TE | | – | – | – | 155–175 | 8–12 |
| –TF | | 525–545 | 4–12 | Hot water | 155–175 | 8–12 |
| LM 26–TE | Al Si9 Cu3 Mg | – | – | – | 200–210 | 7–9 |
| LM 28–TE | Al Si19 Cu Mg Ni | – | – | – | 185 | † |
| –TF | | 495–505 | 4 | Air blast | 185 | 8 |
| LM 29–TE | Al Si23 Cu Mg Ni | – | – | – | 185 | † |
| –TF | | 495–505 | 4 | Air blast | 185 | 8 |
| LM 30–TS | Al Si17 Cu4 Mg | – | – | – | 175–225 | 8 |
| **BS 'L' series** | | | | | | |
| (4L 35) | Al Cu4 Ni2 Mg2 | 500–520 | 6 | Boiling water | 95–110 or room temperature | 2 5 days |
| 3L 51 | Al Si2 Cu Ni Fe Mg | – | – | – | 150–175 | 8–24 |
| (3L 52) | Al Cu2 Ni Si Fe Mg | 520–540 | 4 | Water at 30–100 °C | 150–180 or 195–205 | 8–24 2–5 |
| (4L 53) | Al Mg10 | 425–435 | 8 | Oil at 160 °C max[4] | – | – |
| 3L 78 | Al Si4 Cu1 | 520–530 | 12 | Hot water | 160–170 | 8–10 |
| (2L 91) | Al Cu4 | 525–545 | 12–16 | Hot water | 120–140 | 1–2 |
| 2L 92 | Al Cu4 | 525–545 | 12–16 | Hot water | 120–170 | 12–14 |
| (L 99) | Al Si6 | 535–545 | 12 | Hot water | 150–160 | 4 |
| (L 119) | Al Cu5 Ni1 | 542 ± 5 | 5 | Boiling water or oil at 80 °C | 215 ± 5 | 12–16 |
| L 154 | Al Cu4 Si1 | 510 ± 5 | 16 | Water (50–70 °C) | – | 30 days |
| L 155 | Al Cu4 Si1 | 510 ± 5 | 16 | Water (50–70 °C) | 140 ± 10 | 16 |
| **DTD specifications** | | | | | | |
| 722B | Al Si5 | | | | 165 ± 10 | 8–12 |
| 727B | Al Si5 | 540 ± 5 | 4–12 | Water (80–100 °C) or oil | 130 ± 10 | 1–2 |
| 735B | Al Si5 | 540 ± 5 | 4–12 | Water (80–100 °C) | 165 ± 10 | 8–12 |
| 5008B | Al Zn5 Mg | – | – | – | 180 ± 5 | 10 |
| 5018A | Al Mg7 Zn | 430 ± 5 | 8 | Oil > 160 °C > 1 h then oil at room temperature, or air | – | – |
| or | | 440 ± 5 | 8 | | | |
| then | | 495 ± 5 | 8 | Boiling water | – | – |

[1] Single figures are minimum times at temperature for average castings and may have to be increased for particular castings.
[2] Hot water means water at 70–80 °C unless otherwise stated.
[3] The exact number of hours depends on the mechanical properties required.
[4] The castings may be allowed to cool to 385–395 °C in the furnace before quenching. The castings shall be allowed to stay in the oil for not more than 1 h and may then be quenched in water or cooled in air.
† The duration of the treatment shall be such as will produce the specified Brinell hardness in the castings.
* For temper designation see Table 7.2.
() Specification now withdrawn.

**Table 7.2**   TYPICAL HEAT TREATMENT DATA FOR WROUGHT ALUMINIUM ALLOYS
Times and temperatures within the limits shown. some specifications give tighter limits

| Material designation | Alloy type | Temper‡ | Solution treatment Temperature °C | Quench medium† | Ageing temperature °C | Time at temperature h |
|---|---|---|---|---|---|---|
| 2011 | Al Cu5.5 Pb Bi | T 3 (TD) | 510 ± 5 | Water | Room | 48 |
| | | T 6 (TF) | 510 ± 5 | Water | 155–165 | 12 |

**Table 7.2**   (*continued*)

| Material designation | Alloy type | Temper‡ | Solution treatment Temperature °C | Quench medium† | Ageing temperature °C | Time at temperature h |
|---|---|---|---|---|---|---|
| 2014A | Al Cu4 Si Mg | T 3 (TD) | 505 ± 5 | Water | Room | 48 |
| | | T 4 (TB) | 505 ± 5 | Water | Room | 48 |
| | | T 6 (TF) | 505 ± 5 | Water | 155–190 | 5–20 |
| | | T 651 | 505 ± 5 | Water | 155–190 | 5–20 |
| 2024 | Al Cu4 Mg1 | T 351 | 495 ± 5 | Water | Room | 48 |
| | | T 4 (TB) | 495 ± 5 | Water | Room | 48 |
| | | T 42 | 495 ± 5 | Water | Room | 48 |
| 2031 | Al Cu2 Ni1 Mg Te Si | T 4 (TB) | 525 ± 10 | Water or oil | 155–205 | 2–20 |
| 2117 | Al Cu2 Mg | T 4 (TB) | 495 ± 5 | Water | Room | 96 |
| 2618A | Al Cu2 Mg15 Te1 Ni1 | T 6 (TF) | 530 ± 5 | Water | 160–200 | 16–24 |
| 6061 | Al Mg1 Si Cu | T 4 (TB) | 525 ± 15 | Water | Room | – |
| | | T 6 (TF) | 525 ± 15 | Water | 165–195 | 3–12 |
| 6063 | Al Mg Si | T 4 (TB) | 525 ± 5 | Water | Room | – |
| | | T 5 (TE) | – | – | 160–180 | 5–15 |
| | | T 6 (TF) | 525 ± 5 | Water | 160–180 | 5–15 |
| 6082 | Al Si1 Mg Mn | T 4 (TB) | 530 ± 10 | Water | Room | 1 20 |
| | | T 6 (TF) | 530 ± 10 | Water | 175–185 | 7–12 |
| | | T 651 | 525 ± 15 | Water | 165–195 | 3–12 |
| 6101A | Al Mg Si | T 4 (TB) | 525 ± 5 | Water | Room | 1 20 |
| | | T 6 (TF) | 525 ± 5 | Water | 170 ± 10 | |
| 6463 | Al Mg Si | T 4 (TB) | 525 ± 5 | Water | Room | 1 20 |
| | | T 6 (TF) | 525 ± 5 | Water | 170 ± 10 | 5–15 |
| 7010 | Al Zn6 Mg2 Cu2 | T 351 | 475 ± 10 | Water | – | – |
| | | T 7651 | 475 ± 10 | Water | 172 ± 3* | 6–15 |
| | | | | or | 120 ± 3 | 24 |
| | | | | followed by | 172 ± 3 | 6–15 |
| | | T 73651 | 465 ± 10 | Water | 172 ± 3* | 10–24 |
| 7014 | Al Zn5.5 Mg2 Cu Mn | T 6 (TF) | 460 ± 10 | Water 85 °C or oil | 135 ± 5 | 12 |
| | | T 6510 | 460 ± 10 | Water or oil | 135 ± 5 | 12 |
| 7075 | Al Zn6 Mg Cu | T 6 (TF) | 460 ± 10 | Water | 135 ± 5 | 12 |
| | | T 651 | 460 ± 10 | Water | 135 ± 5 | 12 |
| | | T 73 | 465 ± 5 | Water (60–80 °C) | 110 ± 5 | 6–24 |
| | | | | or | 120 ± 5 | 20–30 |
| | | | | followed by | 177 ± 5 | 6–10 |
| | | T 7351 | 465 ± 5 | Water | 110 ± 5 | 6–24 |
| | | | | followed by | 177 ± 5 | 5–12 |
| | | T 7352 | 465 ± 5 | Water 70 °C | 110 ± 5 | 6–24 |
| | | | | or | 120 ± 5 | 20–30 |
| | | | | followed by | 177 ± 5 | 6–10 |

*Heating to temperature at not more than 20 °C/h.
†Water below 40 °C unless otherwise stated.
‡For temper designation see Table 7.3

**Table 7.3**   ALUMINIUM ALLOY TEMPER DESIGNATIONS

| Symbol | Condition |
|---|---|
| **Casting alloys BS 1490** | |
| M | As cast |
| TB | Solution treated and naturally aged |
| TB7 | Solution treated and stabilized |
| TE | Artificially aged |
| TF | Solution treated and artificially aged |
| TF7 | Solution treated, artificially aged and stabilized |
| TS | Thermally stress relieved |

*continued overleaf*

**Table 7.3**  *(continued)*

| Symbol | Condition |
|---|---|
| **BSEN515** | |
| T 1 | Cooled from elevated temperature shaping process and naturally aged to stable condition |
| T 2 | As T 1 but cold worked after cooling from elevated temperature |
| T 3 (TD) | Solution treated, cold worked and naturally aged to stable condition |
| T 4 (TB) | Solution treated and naturally aged to stable condition |
| T 5 (TF) | Cooled from elevated temperature shaping process and artificially aged |
| T 6 | Solution treated and artificially aged |
| T 7 | Solution treated and stabilized (over-aged) |
| T 8 (TH) | Solution treated, cold worked and then artificially aged |
| T 9 | Solution treated, artificially aged and then cold worked |
| T 10 | Cooled from elevated temperature shaping process artificially aged and then cold worked |

*British equivalents in parenthesis.

## 7.2  Magnesium alloys

### 7.2.1  Safety requirements

A potential fire hazard exists in the heat treatment of magnesium alloys. Overheating and direct access to radiation from heating elements must be avoided and the furnace must be provided with a safety cutout which will turn off heating and blowers if the temperature goes more than 6 °C above the maximum permitted. In a gastight furnace a magnesium fire can be extinguished by introducing boron trifluoride gas through a small opening in the closed furnace after the blowers have been shut down.

### 7.2.2  Environment

For temperature over 400 °C, surface oxidation takes place in air. This can be suppressed by addition of sufficient sulphur dioxide, carbon dioxide or other suitable oxidation inhibitor.

In the case of castings to MEL ZE63A and related specifications, solution treatment should be carried out in an atmosphere of hydrogen and quenching of castings from solution treatment temperature of MEL QE22 is to be done in hot water.

If microscopic examination reveals eutectic melting or high temperature oxidation, rectification cannot be achieved by reheat-treatment. Quench from solution treatment should be rapid, either forced air or water quench. From ageing treatment, air cool.

### 7.2.3  Conditions for heat treatment of magnesium alloys castings

These are shown in Table 7.4 and for some wrought magnesium alloys in Table 7.5. Stress relief treatments are given in Table 7.6.

**Table 7.4**  HEAT TREATMENT OF MAGNESIUM CASTING ALLOYS

| Specifications | Composition | Solution treatment | | Ageing | |
|---|---|---|---|---|---|
| | | Temperature (°C) | Time (h) quench | Temperature (°C) | Time (h) quench |
| MEL ZRE1 | Zn2.5 Zr0.6 | – | – | 250 | 16 AC |
| BS 2L126 | RE3.5 | | | | |
| BS 2970 MAG 6 | | | | | |
| ASTM EZ33A | | | | | |
| UNS M12330 | | | | | |

**Table 7.4** (*continued*)

| Specifications | Composition | Solution treatment | | Ageing | |
|---|---|---|---|---|---|
| | | Temperature (°C) | Time (h) quench | Temperature (°C) | Time (h) quench |
| MEL RZ5 BS 2L128 BS 2970 MAG 5 ASTM ZE41A UNS M16410 | Zn4.2 Zr0.7 RE1.3 | – | – | 330 +170–180 | 2 AC 10–16 AC |
| MEL ZE63A DTD 5045 ASTM ZE63A | Zn5.8 Zr0.7 RE2.5 | 480* | 10–72 WQ | 140 | 48 AC |
| MELZT1 DTD 5005A BS 2970 MAG 8 ASTM HZ32A | Zn2.2 Zr0.7 Th3.0 | – | – | 315 | 16 AC |
| MEL TZ6 DTD 5015A BS 2970 MAG 9 ASTM ZH62A UNS M16620 | Zn5.5 Zr0.7 Th1.8 | – | – | 330 +170–180 | 2 AC 10–16 AC |
| MEL EQ21A | Ag1.5 Zr0.7 Cu0.07 Nd(RE)2.0 | 520 | 8 WQ | 200 | 12–16 AC |
| MEL MSR-B DTD 5035A | Ag2.5 Zr0.6 Nd(RE)2.5 | 520–530 | 4–8 WQ | 200 | 8–16 AC |
| MEL QE22 (MSR) DTD 5055 ASTM QE22A UNS M18220 | Ag2.5 Zr0.6 Nd(RE)2.0 | 520–530 | 4–8 WQ Hot WQ | 200 | 8–16 AC |
| MEL A8 BS 3L122 BS 2970 MAG1 ASTM AZ81A UNS M11818 | Al8.0 Zn0.5 Mn0.3 | 380–390 410–420 | 8 AC 16 AC | – | – |
| MEL AZ91 (ST) BS 3L124 BS 2970 MAG 3 | Al9.0 Zn0.5 Mn0.3 Be0.0015 | 380–390 410–420 | 8 AC 16 AC | – | – |
| ASTM AZ91C (ST&PT) UNS M11914 | Al9.0 Zn0.5 Mn0.3 Be0.0015 | 380–390 410–420 | 8 AC 16 AC | 200 | 10 AC |
| MEL MAG 7 (ST) BS 2970 MAG 7 | Al7.5/9.5 Zn0.3/1.5 Mn0.15 | 380–390 410–420 | 8 AC 16 AC | – | – |
| MEL MAG 7 (ST&PT) | Al7.5/9.5 Zn0.3/1.5 Mn0.15 | 380–390 410–420 | 8 AC 16 AC | 200 | 10 AC |

*In hydrogen. Max 490 °C.

**Table 7.5** HEAT TREATMENT OF MAGNESIUM WROUGHT ALLOYS

| Specifications | Composition | Form | Solution treatment | | Ageing | |
|---|---|---|---|---|---|---|
| | | | Temperature (°C) | Time (h) quench | Temperature (°C) | Time (h) quench |
| MEL AZ80 ASTM AZ80A UNS M11800 | Al8.5 Zn0.5 Mn0.12 | Ex | – | – | 177 | 16 AC |
| | | F | 400 | 2-4 WQ | 177 | 16-24 AC |

*continued overleaf*

**Table 7.5**   *(continued)*

| Specifications | Composition | Form | Solution treatment | | Ageing | |
|---|---|---|---|---|---|---|
| | | | *Temperature* (°C) | *Time* (h) *quench* | *Temperature* (°C) | *Time* (h) *quench* |
| ASTM HM31A | Th2.5–4.0 | Ex | – | – | 232 | 16 AC |
| UNS 13310 | Zn0.3 | | | | | |
| | Zr0.4–1.0 | | | | | |
| ASTM 60A | Zn5.5 | F T6 | 500 | 2 WQ | 150 | 24 AC |
| UNS 16600 | | F T4 | 500 | 2 WQ | – | – |
| | | F T5 | – | – | 150 | 24 AC |

*Notes:* Ex – extrusions, F – forgings, T4 – solution treated, T5 – cooled and artificially aged, T6 – solution treated and artificially aged, AC – air cool, WQ – water quench.

**Table 7.6**   STRESS RELIEF TREATMENTS FOR WROUGHT MAGNESIUM ALLOYS

| Specifications | Composition | Form | *Temperature* (°C) | *Time* (min) |
|---|---|---|---|---|
| MEL AZM | Al6.0 Zn1.0 | Ex&F | 260 | 15 |
| ASTM Al61A | Mn0.3 | SH | 204 | 60 |
| UNS 11610 | | SA | 343 | 120 |
| MEL AZ80 | Al8.5 Zn0.5 | Ex&F | 260 | 15 |
| ASTM AZ80 | Mn0.12 min | Ex&F* | 204 | 60 |
| UNS 11311 | | | | |
| MEL AZ31 | Al3.0 Zn1.0 | Ex&F | 260 | 15 |
| ASTM AZ31B | Mn0.3 | SH | 149 | 60 |
| UNS 11311 | | SA | 343 | 120 |

*Notes:* Ex – extrusions, F – forgings, SH – sheet hard rolled, SA – sheet annealed, * – cooled and artificially aged.

# 8 Metal finishing

The processes and solutions described in this section are intended to give a general guide to surface finishing procedures. To operate these systems on an industrial scale would normally require recourse to one of the Chemical Supply Houses which retail properietary solutions.

## 8.1 Cleaning and pickling processes

VAPOUR DEGREASING

Used to remove excess oil and grease. Components are suspended in a solvent vapour, such as tri- or tetrachloroethylene.
*Note*: Both vapours are toxic and care should be taken to ensure efficient condensation or extraction of vapours.

EMULSION CLEANING

An emulsion cleaner suitable for most metals can be prepared by diluting the mixture given below with a mixture of equal parts of white spirit and solvent naphtha.

| | |
|---|---|
| Pine oil | 62 g |
| Oleic acid | 10.8 g |
| Triethanolamine | 7.2 g |
| Ethylene glycol-monobutyl ether | 20 g |

This is used at room temperature and should be followed by thorough swilling.

**Table 8.1** ALKALINE CLEANING SOLUTIONS

| Metal to be cleaned | Composition of solution | | | Temperature | | Remarks |
|---|---|---|---|---|---|---|
| | | oz gal$^{-1}$ | g l$^{-1}$ | °F | °C | |
| All common metals other than aluminium and zinc, but including magnesium | Sodium hydroxide (NaOH) | 6 | 37.5 | 180–200 | 80–90 | For heavy duty |
| | Sodium carbonate (Na$_2$CO$_3$) | 4 | 25.0 | | | |
| | Tribasic sodium phosphate (Na$_3$PO$_4$.12H$_2$O) | 1 | 6.2 | | | |
| | Wetting agent | $\frac{1}{4}$ | 1.5 | | | |

*continued overleaf*

**Table 8.1**   (*continued*)

| Metal to be cleaned | Composition of solution | | | Temperature | | Remarks |
|---|---|---|---|---|---|---|
| | | oz gal$^{-1}$ | g l$^{-1}$ | °F | °C | |
| | Sodium hydroxide | 2 | 12.5 | 180–200 | 80–90 | For medium duty |
| | Sodium carbonate | 4 | 25.0 | | | |
| | Tribasic sodium phosphate | 2 | 12.5 | | | |
| | Sodium metasilicate (Na$_2$SiO$_3$.5H$_2$O) | 2 | 12.5 | | | |
| | Wetting agent | $\frac{1}{8}$ | 0.75 | | | |
| | Tribasic sodium phosphate | 4 | 25.0 | 180–200 | 80–90 | For light duty |
| | Sodium metasilicate | 4 | 25.0 | | | |
| | Wetting agent | $\frac{1}{8}$ | 0.75 | | | |
| Aluminium and zinc | Tribasic sodium phosphate | 2 | 12.5 | 180–200 | 80–90 | – |
| | Sodium metasilicate | 4 | 25.0 | | | |
| | Wetting agent | $\frac{1}{8}$ | 0.75 | | | |
| Most common metals | Sodium carbonate | 2 | 12.5 | 180–200 | 80–90 | Electrolytic cleaner, 6 V Current density |
| | Tribasic sodium phosphate | 4 | 25.0 | | | 100/A ft$^{-2}$ (10/A dm$^{-2}$) |
| | Wetting agent | $\frac{1}{4}$ | 1.5 | | | Article to be cleaned may be made cathode or anode or both alternately |
| Most common metals | Sodium carbonate | 6 | 37.5 | Room | Room | May be used electrolytically |
| | Sodium hydroxide | 1 | 6.25 | | | |
| | Tribasic sodium phosphate | 2 | 12.5 | | | |
| | Sodium cyanide (NaCN) | 2 | 12.5 | | | |
| | Sodium metasilicate | 1 | 6.25 | | | |
| | Wetting agent | $\frac{1}{8}$ | 0.75 | | | |

**Table 8.2**   PICKLING SOLUTIONS

| Metal to be pickled | Composition of solution | | | Temperature | | Remarks |
|---|---|---|---|---|---|---|
| | | oz gal$^{-1}$ | g l$^{-1}$ | °F | °C | |
| Aluminium (wrought) | *For etching* Sodium hydroxide (NaOH) | 8 | 56 | 104–176 | 40–80 | Articles dipped until they gas freely, then swilled, and dipped in nitric acid 1 part by vol. to 1 of water (room temperature) |
| Aluminium (cast and wrought) | Nitric acid, s.g. 1.42 | 1 gal | 11 | Room | Room | Articles first cleaned in solvent degreaser. Use polythene or PVC tanks |
| | Hydrofluoric acid (52%) | 1 gal | 11 | | | |
| | Water | 8 gal | 81 | | | |

**Table 8.2** (*continued*)

| Metal to be pickled | Composition of solution | | | Temperature | | Remarks |
|---|---|---|---|---|---|---|
| | | oz gal$^{-1}$ | g l$^{-1}$ | °F | °C | |
| | *Bright dip* | | | | | |
| | Chromic acid | 0.84 oz | 5.2 g | 195 | 90 | Immerse for $1\frac{1}{2}$ min. |
| | Ammonium bifluoride | 0.72 oz | 4.5 g | | | Solution has limited life. |
| | Cane syrup | 0.68 oz | 4.2 g | | | AR chemicals and |
| | Copper nitrate | 0.04 oz | 0.25 g | | | deionized or distilled |
| | Nitric acid (s.g. 1.4) | 4.8 oz | 30 ml | | | water should be used |
| | Water (distilled) to | 1 gal | 11 | | | |
| Aluminium and other non-ferrous metals | *Bright dip* Phosphoric acid (s.g. 1.69) | 8.4 gal | 9.41 | 195 | 90 | Immerse for several |
| | Nitric acid (s.g. 1.37) | 0.6 gal | 0.61 | | | min. Agitate work and solution. Good ventilation necessary. Addition of acetic acid |
| Magnesium and magnesium alloys | *General cleaner* Chromic acid | 16–32 | 100–200 | Up to b.p. | Up to b.p. | For removal of oxide films, corrosion products, etc. Should not be used on oily or painted material |
| | *Sulphuric acid pickle* Sulphuric acid* | 3% | – | Room | Room | Should be used on rough castings or heavy sheet only. Removes approx. 0.002 in. in 20–30 s |
| | *Nitro-sulphuric pickle* Nitric acid | 8% | – | Room | Room | |
| | Sulphuric acid* | 2% | – | | | |
| | *Bright pickle for wrought products* Chromic acid | 23 | 150 | Room | Room | Lustrous appearance. |
| | Sodium nitrate | 4 | 25 | | | Involves metal removal |
| | Calcium or magnesium fluoride | $\frac{1}{8}$ | $\frac{3}{4}$ | | | |
| | *Bright pickle for castings* Chromic acid | $37\frac{1}{2}$ | 235 | Room | Room | |
| | Concentrated nitric acid (70%) | $3\frac{1}{4}$ | 20 | | | |
| | Hydrofluoric acid (50%) | 1 | 6.2 | | | |
| | *Acetic acid* | 8 approx. | 50 approx. | Room | Room | Special purpose pickles |
| | *Citric acid* | 8 approx. | 50 approx. | Room | Room | Special purpose pickles |

*Note*: It is almost universal practice to use an inhibitor in the pickling bath. This ensures dissolution of the scale with practically no attack on the metal. Inhibitors are usually of the long chain amine type and often proprietary materials. Examples are Galvene and Stannine made by ICI.

## 8.2   Anodizing and plating processes

**Table 8.3**   ANODIZING PROCESSES FOR ALUMINIUM
Good ventilation above the bath and agitation of the bath is advisable in all cases.

| | Composition of solution | | Temperature | | Current density amp ft⁻² | Time and voltage | Cathodes | Vat | Hangers | Remarks |
|---|---|---|---|---|---|---|---|---|---|---|
| | oz gal⁻¹ | g l⁻¹ | °F | °C | (A dm⁻²) | | | | | |
| Chromic acid (CrO₃), chloride content must not exceed 0.2 g l⁻¹ sulphate less than 0.5 g l⁻¹ (After Bengough- Stuart) | 5–16 | 30–100 | 103–108 | 38 42 | Current controlled by voltage. Average 3–4 (0.3–0.4) d.c. | †1–10 min 0–40 V increased in steps of 5 V 5–35 min Maintain at 40 V 3–5 min Increase gradually to 50 V 4–5 min Maintain at 50 V | Tank or stainless steel | Steel (exhausted) | Pure aluminium or titanium | Slight agitation is required. This process cannot be used with alloys containing more than 5% copper |
| Sulphuric acid (s.g. 1.84) | 32 | 200 | 60–75 | 15–24 | 10–20 (1–2) d.c. | 12–18 V 20–40 min | Aluminium or lead plates (tank if lead lined) | Lead lined steel | Pure aluminium or titanium | The current must not exceed 0.2 A l⁻¹ of electrolyte |
| Hard anodizing Hardas process Sulphuric acid | 32 | 200 | 23–41 | −5– + 5 | 25–400 (2.5–40) d.c. | 40–120 V | Lead | Lead lined steel | Aluminium or titanium | Agitation required. Gives coating 1–3 thou. thick |
| Eloxal GX process Oxalic acid (COOH)₂.2H₂O | 12.8 | 80 | 70 | 20 | 10–20 (1–2) d.c. | 50 V 30–60 min | Vat lining | Lead lined steel | Aluminium or titanium | Oxalic acid processes are more expensive than sulphuric acid anodizing; but coatings are thicker and are coloured. |

| | | | | | | | | | |
|---|---|---|---|---|---|---|---|---|---|
| *Eloxal WX process*<br>Oxalic acid | 12.8 | 80 | 75–95 | 25–35 | 20–30<br>(2–3) a.c. | 20–60 V<br>40–60 min | Vat lining | Lead lined steel | Aluminium or titanium |
| *Integral colour*<br>*Anodizing Kalcolor*<br>*process* | 0.8 | 5 | | | | | | | |
| Sulphuric acid | 16 | 100 | 72 | 22 | 30<br>(3) d.c. | 25–60 V<br>20–45 min | Lead | Lead lined steel | Aluminium or titanium<br>Aluminium level in solution must be maintained between 1.5 and 3 g l$^{-1}$ |
| Sulphosalicylic acid | | | | | | | | | |

† Period according to degree of protection Complete cycle normally 40 min.

**Table 8.4** ANODIZING PROCESSES FOR MAGNESIUM ALLOYS

| Composition of solution | oz gal⁻¹ | g l⁻¹ | Temperature F | Temperature C | Current density A ft⁻² (A dm⁻²) | Time and voltage | Cathodes | Vat | Hangers | Remarks |
|---|---|---|---|---|---|---|---|---|---|---|
| *HAE process* | | | <95 | <35 | 12–15 (1.2–1.5) | 90 min at 85 V approx. a.c. preferred | Mg alloy for a.c. Mg or steel if d.c. used | Mild steel or rubber lined | Mg alloy | Matt hard, brittle, corrosion resistant, dark brown 25–50 $\mu$m thick, abrasion resistant |
| Potassium hydroxide | 19.2 | 120 | | | | | | | | |
| Aluminium | 1.7 | 10.4 | | | | | | | | |
| Potassium fluoride | 5.5 | 34 | | | | | | | | |
| Trisodium phosphate | 5.5 | 34 | | | | | | | | |
| Potassium manganate | 3.2 | 20 | | | | | | | | |
| *Dow 17 process* | | | 160–180 | 70–85 | 5–50 (0.5–5) | 10–100 min up to 110 V a.c. or d.c. | Mg alloy for a.c. Mg or steel for d.c. | Mild steel or rubber lined | Mg alloy | Matt dark green, corrosion resistant, 25 $\mu$m thick approx., abrasion resistant |
| Ammonium bifluoride | 39 | 232 | | | | | | | | |
| Sodium dichromate | 16 | 100 | | | | | | | | |
| Phosphoric acid 85% $H_3PO_4$ | 14 | 88 | | | | | | | | |
| *Cr 22 process* | | | 165–205 | 75–95 | 15 (1.5) | 12 min 380 V a.c. | – | Mild steel | Mg alloy | Matt dark green, corrosion resistant, 25 $\mu$m thick approx. |
| Chromic acid | 4 | 25 | | | | | | | | |
| Hydrofluoric acid (50%) | 4 | 25 | | | | | | | | |
| Phosphoric acid $H_3PO_4$ (85%) | 13.5 | 84 | | | | | | | | |
| Ammonia solution | 25–30 | 160–180 | | | | | | | | |
| *MEL process* Fluoride anodize | | | <86 | <30 | 5–100 (0.5–10) | 30 min 120 V a.c. preferred | Mg alloy for a.c. Mg or steel for d.c. | Rubber lined | Mg alloy | Principally a cleaning process to improve corrosion resistance by dissolving or ejecting cathodic particles from the surface |
| Ammonium bifluoride | 16 | 100 | | | | | | | | |

## 8.3   Plating processes for magnesium alloys

DOW PROCESS (H. K. DELONG)

This process depends on the formation of a zinc immersion coat in a bath of the following composition:

| Component | Concentration oz gal$^{-1}$ | g l$^{-1}$ |
|---|---|---|
| Tetrasodium pyrophosphate | 16 | 120 |
| Zinc sulphate | 5.3 | 40 |
| Potassium fluoride | 1.0 | 7 |

The treatment time is 3–5 min at a temperature of 175–185 F (80–85 °C) with mild agitation. The pH of the bath should be 10–10.4.
The steps of the complete process are:

1. Solvent or vapour degreasing.
2. Hot caustic soda clean or cathodic cleaning in alkaline cleaner.
3. Pickle $\frac{1}{4}$–$1\frac{1}{2}$ min in 1% hydrochloric acid and rinse.
4. Zinc immersion bath as above without drying off from the rinse.
5. Cold rinse and immediately apply copper strike as under.

| Component | Concentration oz gal$^{-1}$ | g l$^{-1}$ |
|---|---|---|
| Copper cyanide | 4.2 | 26 |
| Potassium cyanide | 7.4 | 46 |
| Potassium carbonate | 2.4 | 15 |
| Potassium hydroxide | 1.2 | 7.5 |
| Potassium fluoride | 4.8 | 30 |
| Free cyanide | 1.2 | 7.5 |
| pH | 12.8–13.2 | – |
| Temperature | 140 °F (60 °C) | – |

CONDITIONS

30–40 A ft$^{-2}$ (3–4 A dm$^{-2}$) for $\frac{1}{2}$–1 min, reducing to 15–20 A ft$^{-2}$ (1.5–2 A dm$^{-2}$) for 5 min or longer.
If required, the copper thickness from the above strike can be built up in the usual alkaline or proprietary bright plating baths. Following the above steps, further plating may be carried out in conventional electroplating baths.

ELECTROLESS PLATING ON MAGNESIUM

Deposits of a compound of nickel and phosphorus can be obtained on magnesium alloy components by direct immersion in baths of suitable compositions. Details of the process may be obtained from the inventors. The Dow Chemical Co. Inc., Midland, Michigan, USA.

'GAS PLATING' OF MAGNESIUM (VAPOUR PLATING)

Deposits of various metals on magnesium components (as on other metals) can be produced by heating the article in an atmosphere of a carbonyl or hydride of the metal in question.

# 9 Superplasticity of light metal alloys

Superplasticity is the name given to the ability of a material to sustain extremely large deformations at low flow stresses at a temperature around half the melting point expressed in Kelvin. It is only found in metals and alloys, which have, and can maintain during forming, a very fine grain structure. A parameter which indicates the degree of superplasticity is the strain rate sensitivity $m$, given by the high temperature flow equation: $\sigma = K\dot{e}^m$, $\sigma$ is the stress for plastic flow, $\dot{e}$ the applied strain rate and $K$ is a constant. Superplastic materials have $m$ values normally between 0.4 and 0.6, while most other metals and alloys at elevated temperatures have $m$ values of 0.2; viscous materials (e.g. glass) behave like a Newtonian fluid and have $m$ values of 1.

A full discussion of the mechanism of superplasticity, including methods for determining $m$, can be found in K. A. Padmanabhan and G. J. Davies, 'Superplasticity', Berlin, Springer-Verlag, 1980.

The tables in this chapter give alloy systems with the temperature range over which they show superplasticity, the maximum possible percentage elongation, and the $m$ value. The values of about $10^{-4}$ quoted under remarks are preferred strain rates.

**Table 9.1**  NON-FERROUS-SYSTEMS (LIGHT METAL ALLOY)

| Alloy system | Temperature range °C | Maximum elongation % | m | Remarks |
|---|---|---|---|---|
| Al (commercial) | 380–580 | 6 000 | 0.2 | |
| Al–7.6Ca | 400–600 | 850 | 0.78 | Euratom alloy |
| Al–7.6Ca | ~500 | 570 | 0.32 | Optimum 500 °C at $4.16 \times 10^{-4}$ |
| Al–5Ca–5Zn | 450 | | | Alcan 08050 |
| Al–17Cu | 400–520 | 600 | 0.35 | |
| Al–32:74Cu | ~510 | >160 | 0.7 | After 25% pre-deformation at 510 °C |
| Al–33Cu | 380–520 | 1 150 | 0.9 | |
| Al–6Cu–0.5Zr | 400–500 | 2 000 | 0.5 | TI Supral |
| Al–2.59Cu–2.26Li–0.16Zr(2090) | 450–480 | 1 800 | – | Optimum 520 °C at $5 \times 10^{-4}$ |
| Al–4.40Cu–0.70Mg–0.80Si–0.75Mn(2014) + 15%SiC | 450–480 | 160 | 0.4 | Elong% 350% at 5.25 MPa hydrostatic pressure |
| Al–25–33Cu–7–11Mg | 420–480 | >600 | 0.72 | |
| Al–Ga–Ti | RT | – | 0.3–0.5 | |
| Al–4Ge | 400–500 | 230 | – | |
| Al–2.4Li–2.0Cu–0.70Mg–0.08Zr(8091 | 490–550 | 1 350 | 0.6 | Optimum 530 °C at $10^{-3} - 10^{-4}$ |
| Al–2.5Li–1.2Cu–0.6Mg–0.1Zr(8090) | 520–530 | 660–1 200 | 0.65 | 2–3 mm sheet at $2-5 \times 10^{-4}$ |

**Table 9.1** *(continued)*

| Alloy system | Temperature range °C | Maximum elongation % | m | Remarks |
|---|---|---|---|---|
| Al−2.4Li−1.2Cu−0.60Mg−0.12Zr(8090) | ~530 | >1 000 | 0.68 | 2 mm sheet at $5 \times 10^{-4}$ and 5−10 micron grain size |
| Al−1.56Mg−5.6Zn | 530 | 500 | 0.7 | |
| Al−3Mg−6Zn | 340−360 | 400 | 0.35 | TI BA 480 |
| Al−0.93Mg−10.72Zn−0.42Zr | 550 | 1 550 | 0.9 | |
| Al−5.8Mg−0.37Zr + others | 520 | >800 | 0.6 | |
| Al−4.89 Mg−1.19Cr | 482−520 | >1 000 | | Optimum 520 °C at $1.6 \times 10^{-2}$ |
| Al−8.0Mg−1.0Li−0.15Zr | 300 | >1 000 | | |
| Al−5.70Zn−2.30Mg−1.50Cu−0.22Cr(7475) + 15% SiC | 495−515 | 97 | 0.43 | Elong% 310% at 5.25 MPa hydrostatic pressure |
| Al−5.70Zn−2.30Mg−1.50Cu−0.22Cr(7475) | ~530 | 1 300 | | Optimum 530 °C at $2.8 \times 10^{-4}$ |
| Mg−33Al | 350−400 | 2 100 | 0.8 | Eutectic |
| Mg−9Li | 180 | 460 | 0.52 | At $3 \times 10^{-4}$; 6.1 micron grain size |
| Mg−9Li | 200 | 445 | 0.47 | At $1 \times 10^{-3}$; 7.1 micron grain size |
| Mg−9Li | 250 | 310 | 0.44 | At $1 \times 10^{-4}$; 14.2 micron grain size |
| Mg−4.3Al−3Zn−0.5Mn | | | | Russian MA 15 |
| Mg−30.7Cd | 450 | 250 | − | |
| Mg−5.5Zn−0.5Zr | 270−310 | 1 000 | 0.6 | ZK 60 |
| Mg−0.5Zr | 500 | 150 | 0.3 | |
| Ti (commercial) | 900 | − | 0.8 | RC 70 |
| Ti−4Al−0.25O₂ | 950−1 050 | − | 0.6 | |
| Ti−5Al−2.5Sn | 900−1 100 | 450 | 0.72 | |
| Ti−6Al−4V | 750−1 000 | 1 000 | 0.85 | Commercial alloy used throughout world |
| Ti−6Al−5Zr−4Mo−1Cu−0.25Si | 800 | 300 | − | IMI 700 |
| Ti−8Mn | 580−900 | 140 | 0.95 | |
| Ti−15Mo | 580−900 | 450 | 0.6 | |
| Ti−11Sn−5Zr−2.25Al−1Mo−0.25Si | 800 | 500 | − | IMI 679 |

# 10   Light metal-matrix composites

Metal-matrix composites are engineered materials comprising reinforceants of high elastic modulus and high strength in a matrix of a more ductile and tougher metal of lower elastic modulus and strength. The metal-matrix composite has a better combination of properties than can be achieved by either component material by itself. The objective of adding the reinforceant is to transfer the load from the matrix to the reinforceant so that the strength and elastic modulus of the composite are increased in proportion to the strength, modulus and volume fraction of the added material.

The reinforcement can take one of several forms. The least expensive and most readily available on the market are the particulates. These can be round but are usually irregular particles of ceramics, of which SiC and $Al_2O_3$ are most frequently used. Composites reinforced by particulates are isotropic in properties but do not make best use of the reinforceant. Fine fibres are much more effective though usually more costly to use. Most effective in load transfer are long parallel continuous fibres. Somewhat less effective are short parallel fibres. Long fibres give high axial strength and stiffness, low coefficients of thermal expansion and, in appropriate matrices, high creep strength. These properties are very anisotropic and the composites can be weak and brittle in directions normal to that of the fibres. Where high two-dimensional properties are needed, cross-ply or interwoven fibres can be used. Short or long randomly oriented fibres provide lower efficiencies in strengthening (but are still more effective than particulates). These are most frequently available as SiC whiskers or as short random alumina ('Saffil') fibres or alumino-silicate matts.

Long continuous fibres include drawn metallic wires, mono-filaments deposited by CVD or multi-filaments made by pyrolysis of polymers. The properties of some typical fibres are compared in Table 10.1. The relative prices are given as a very approximate guide.

Because most composites are engineered materials, the matrix and the reinforceant are not in thermodynamic equilibrium and so at a high enough temperature, reaction will occur between them which can degrade the properties of the fibre in particular and reduce strength and more especially fatigue resistance. As many composites are manufactured by infiltration of the liquid metal matrix into the pack of fibres, reaction may occur at this stage. Some typical examples of interaction are listed in Table 10.2.

In order to obtain load transfer in service, it is essential to ensure that the reinforceant is fully wetted by the matrix during manufacture. In many cases, this requires that the fibre is coated with a thin interlayer which is compatible with both fibre and matrix. In many cases, this also has the advantage of preventing deleterious inter-diffusion between the two component materials. The data on most coatings are proprietary knowledge. However, it is well known that silicon carbide is used as an interlayer on boron and on carbon fibres to aid wetting by aluminium alloys.

The routes for manufacturing composites are still being developed but the most successful and lowest cost so far is by mixing particulates in molten metal and casting to either foundry ingot or as billets for extrusion or rolling. This is applied commercially to aluminium alloy composites. Another practicable route is co-spraying in which SiC particles are injected into an atomized stream of aluminium alloy and both are collected on a substrate as a co-deposited billet which can then be processed conventionally. This is a development of the Osprey process and can be applied more widely to aluminium and other alloys. Other routes involve the infiltration of molten metal into fibre pre-forms of the required shape often contained within a mould to ensure the correct final shape. This can be done by squeeze casting or by infiltrating semi-solid alloys to minimize interaction between the fibre and metal. Fibres can also be drawn through a melt to coat them and then be consolidated by hot-pressing.

**Table 10.1** PROPERTIES OF REINFORCING FIBRES AT ROOM TEMPERATURE

| Fibre | Form | Preparation route | Diameter μm | Density g cm$^{-3}$ | Fracture stress MPa | Elastic modulus GPa | Coefficient of thermal expansion K$^{-1}$ × 10$^6$ | Price relative to glass fibre |
|---|---|---|---|---|---|---|---|---|
| SiC | Cont. mono-filament | Chemical vapour depos. | 150 | 3.4 | 3 800 | 450 | 4.5 | 500 |
| SiC | Cont. multi-filament | Polymer fibre pyrolysis | 10–15 | 2.6 | 2 500 | 200 | 4.5 | 100 |
| SiC | Whisker (random, short) | Polymer fibre pyrolysis | 0.1–2.0 | 3.2 | 10 000 | 700 | 4.5 | 150 |
| ~Al$_2$O$_3$ | Multi-filament | Oxide/salt fibre pyrolysis | 15–25 | 3.9 | 1 500 | 380 | 7.0 | 100 |
| ~Al$_2$O$_3$ | Random short fibres | pyrolysis | 2–4 | 3.5 | 2 000 | 300 | 7.0 | 25 |
| C(high modulus) | Cont. multi-filament | Polymer fibre pyrolysis | 10 | 2.0 | 3 000 | 600 | 0 | 1000 |
| C(med. strength) | Cont. multi-filament | Polymer fibre pyrolysis | 8 | 1.9 | 4 200 | 300 | 0 | 100 |

**Table 10.2**   TYPICAL INTERACTIONS IN SOME FIBRE-MATRIX SYSTEMS

| System | Potential interaction | Temperature of significant interaction °C |
|---|---|---|
| Al–C | Formation of $Al_4C_3$ at interface. | 550 |
| | Degradation of C fibre properties. | ~495 |
| Al–$Al_2O_3$ | No significant reaction at normal fabrication temperatures | |
| Al–oxide ($Al_2O_3$–$SiO_3$–$B_2O_3$) | $B_2O_3$ reacts with Al to form borides. | 770 |
| Al–B | Boride formation; interlayer of SiC needed. | 500 |
| Al/Li–$Al_2O_3$ | Interfacial layer of $LiAl_5O_8$ on liquid infiltration. | ~650 |
| Al–SiC | No significant reaction below melting point. | m.p. 660 |
| | $Al_4C_3$ and Si can form in liquid Al. | >700 |
| Al–steel | Formation of iron aluminides. | 500 |
| Mg(AZ91)–C | No significant reaction at m.p. of alloy provided O and N avoided during infiltration. | |
| Ti–B | Formation of $TiB_2$. | 750 |
| Ti–SiC | Formation of TiC, $TiSi_2$ and $Ti_5Si_3$. | 700 |

**Table 10.3** MECHANICAL PROPERTIES OF ALUMINIUM ALLOY COMPOSITES AT ROOM TEMPERATURE

| Base Alloy | Nominal composition | Form | Heat treatment | % particulate | 0.2% proof stress MPa | Tensile stress MPa | Elongation % | Elastic modulus GPa | Fracture toughness MPa m$^{-1/2}$ | Density g cm$^{-3}$ |
|---|---|---|---|---|---|---|---|---|---|---|
| 6061 | Mg 1.0 Si 0.6 Cu 0.2 Cr 0.25 | Extrusion | T6 | Nil | 276 | 310 | 20.0 | 69.9 | 29.7 | 2.71 |
| | | | | 10% Al$_2$O$_3$ | 297 | 338 | 7.6 | 81.4 | 24.1 | 2.81 |
| | | | | 15% Al$_2$O$_3$ | 317 | 359 | 5.4 | 87.6 | 22 | 2.86 |
| | | | | 20% Al$_2$O$_3$ | 359 | 379 | 2.1 | 98.6 | 21.5 | 2.94 |
| | | | | 13% SiC | 317 | 356 | 4.9 | 89.5 | 17.9 | – |
| | | | | 20% SiC | 440 | 585 | 4.0 | 120.0 | – | – |
| | | | | 30% SiC | 570 | 795 | 2.0 | 140.0 | – | – |
| 2014A | Cu 4.4 Mg 0.7 Si 0.8 Mn 0.75 | Extrusion | T6 | Nil | 414 | 483 | 13.0 | 73.1 | 25.3 | 2.80 |
| | | | | 10% Al$_2$O$_3$ | 483 | 517 | 3.3 | 84.1 | 18.0 | 2.92 |
| | | | | 15% Al$_2$O$_3$ | 476 | 503 | 2.3 | 91.7 | 18.8 | 2.97 |
| | | | | 20% Al$_2$O$_3$ | 483 | 503 | 0.9 | 101.4 | – | 2.98 |
| | | | | 10% SiC | 457 | 508 | 1.8 | 91.2 | 17.7 | – |
| 2219 | Cu 6.0 Mn 0.3 V 0.1 | Sheet | T6 | 8.2% SiC | 448 | 516 | 4.5 | 82.5 | – | – |
| | | Extrusion | T6 | Nil | 290 | 414 | 10.0 | 73.1 | – | – |
| | | | | 15% Al$_2$O$_3$ | 359 | 428 | 3.8 | 88.3 | – | – |
| | | | | 20% Al$_2$O$_3$ | 359 | 421 | 3.1 | 91.7 | – | – |
| 2618 | Cu 2.0 Mg 1.5 Si 0.9 Fe 0.9 Ni 1.0 | Sheet | T6 | ~10% SiC | 396 | 468 | 3.3 | 93.6 | – | – |
| | | Extrusion | T6 | Nil | 320 | 400 | – | 75.0 | – | – |
| | | Extrusion | T6 | 13% SiC | 333 | 450 | 6.0 | 89.0 | 28.9 | – |
| 7075 | Zn 5.6 Mg 2.2 Cu 1.5 Cr 0.2 | Extrusion | T6 | Nil | 617 | 659 | 11.3 | 71.1 | – | – |
| | | | | 12% SiC | 597 | 646 | 2.6 | 92.2 | – | – |
| 8090 | Li 2.5 Cu 1.3 Mg 0.95 Zr 0.1 | Extrusion (18 mm) | T6 | Nil | 480 | 550 | 5.0 | 79.5 | – | – |
| | | | T6 | 12% SiC | 486 | 529 | 2.6 | 100.1 | – | – |

**Table 10.4**    MECHANICAL PROPERTIES OF ALUMINIUM ALLOY COMPOSITES AT ELEVATED TEMPERATURES

| Base Alloy | Nominal composition | | Form | Heat treatment | % particulate | Temperature °C | 0.2% proof stress MPa | Tensile strength MPa |
|---|---|---|---|---|---|---|---|---|
| 6061 | Mg | 1.0 | Extrusion | T6 | 15% $Al_2O_3$ | 22 | 317 | 359 |
| | Si | 0.6 | | | 15% $Al_2O_3$ | 93 | 290 | 331 |
| | Cu | 0.2 | | | 15% $Al_2O_3$ | 150 | 269 | 303 |
| | Cr | 0.25 | | | 15% $Al_2O_3$ | 204 | 241 | 262 |
| | | | | | 15% $Al_2O_3$ | 260 | 172 | 179 |
| | | | | | 15% $Al_2O_3$ | 316 | 110 | 117 |
| | | | | | 15% $Al_2O_3$ | 371 | 62 | 69 |
| 2014A | Cu | 4.4 | Extrusion | T6 | 15% $Al_2O_3$ | 22 | 476(413) | 503(483) |
| | Mg | 0.7 | | | 15% $Al_2O_3$ | 93 | 455(393) | 490(434) |
| | Si | 0.8 | | | 15% $Al_2O_3$ | 150 | 407(352) | 434(379) |
| | Mn | 0.75 | | | 15% $Al_2O_3$ | 204 | 317(283) | 338(310) |
| | | | | | 15% $Al_2O_3$ | 260 | 200(159) | 214(172) |
| | | | | | 15% $Al_2O_3$ | 316 | 103(62) | 110(76) |
| | | | | | 15% $Al_2O_3$ | 371 | 55(35) | 55(41) |

Figures in parentheses are for basic alloy without particulate.

**Table 10.5**    MECHANICAL PROPERTIES OF MAGNESIUM ALLOY COMPOSITES AT ROOM TEMPERATURE

| Base Alloy | Nominal composition | Form | % reinforcement | 0.2% proof stress MPa | Tensile strength MPa | Elongation % | Elastic modulus GPa |
|---|---|---|---|---|---|---|---|
| ZK60A | Mg–5.5Zn–0.5Zr | Extruded rod | Nil | 260 | 325 | 15.0 | 44 |
| | | | 15% SiC(partic.) | 330 | 420 | 4.7 | 68 |
| | | | 20% SiC(partic.) | 370 | 455 | 3.9 | 74 |
| | | | 15% SiC(whisker) | 450 | 570 | 2.0 | 83 |
| | | | 20% $B_4C$(partic.) | 405 | 490 | 2.0 | 83 |
| | Mg–12Li | Squeeze infiltration | Nil | – | 80 | 8.0 | – |
| | | Squeeze infiltration | 12% $Al_2O_3$(fibre) | – | 200 | 3.5 | – |
| | | Squeeze infiltration | 24% $Al_2O_3$(fibre) | – | 280 | 2.0 | – |
| | | Extruded rod | Nil | – | 75 | 10.0 | 45 |
| | | Extruded rod | 20% SiC(whisker) | – | 338 | 0.8 | 112 |

**Table 10.6**    MECHANICAL PROPERTIES OF TITANIUM ALLOY COMPOSITES

| Base Alloy | Form | % particulate | Temperature °C | 0.2% proof stress MPa | Tensile strength MPa | Elongation % | Elastic modulus GPa |
|---|---|---|---|---|---|---|---|
| Ti-6Al-4V | Forging | 10% TiC | 21 | 800 | 806 | 1.13 | 106–120 |
| | | 10% TiC | 427 | 476(393) | 524(510) | 1.70(11.6) | – |
| | | 10% TiC | 538 | 414(359) | 455(441) | 2.40(8.5) | – |
| | | 10% TiC | 649 | 369(221) | 317(310) | 2.90(4.2) | – |
| | | 10% $B_4C$ | 21 | – | 1055(890) | – | 205 |

Figures in parentheses are for basic alloy without particulate.

# Index